T0338041

# PHOSPHORUS POLLUTION CONTROL - POLICIES AND STRATEGIES

# New Analytical Methods in Earth and Environmental Science Series

Because of the plethora of analytical techniques now available, and the acceleration of technological advance, many earth scientists find it difficult to know where to turn for reliable information on the latest tools at their disposal, and may lack the expertise to assess the relative strengths or potential limitations of a particular technique. This new series will address these difficulties, and by providing comprehensive and up-to-date coverage, will rapidly become established as a trusted resource for researchers, advanced students and applied earth scientists wishing to familiarize themselves with emerging techniques in their field.

Volumes in the series will deal with:

- the elucidation and evaluation of new analytical, numerical modelling, imaging or measurement tools/techniques that are expected to have, or are already having, a major impact on the subject;
- new applications of established techniques; and
- Interdisciplinary applications using novel combinations of techniques.

See below for our full list of books from the series:

*Boron Proxies in Paleoceanography and Paleoclimatology*
Bärbel Hönisch, Stephen Eggins, Laura Haynes, Katherine Allen, Kate Holland, Katja Lorbacher

*Structure from Motion in the Geosciences*
Jonathan L. Carrivick, Mark W. Smith, Duncan J. Quincey

*Ground-penetrating Radar for Geoarchaeology*
Lawrence B. Conyers

*Rock Magnetic Cyclostratigraphy*
Kenneth P. Kodama, Linda A. Hinnov

*Techniques for Virtual Palaeontology*
Mark Sutton, Imran Rahman, Russell Garwood

# PHOSPHORUS POLLUTION CONTROL - POLICIES AND STRATEGIES

## ALEKSANDRA DRIZO

CEO, Water and Soil Solutions International Ltd
Edinburgh, UK

**WILEY** Blackwell

*Registered Office(s)*

John Wiley & Sons, Inc., 111 River Street, Hoboken, NJ 07030, USA

John Wiley & Sons Ltd, The Atrium, Southern Gate, Chichester, West Sussex, PO19 8SQ, UK

*Editorial Office*

9600 Garsington Road, Oxford, OX4 2DQ, UK

For details of our global editorial offices, customer services, and more information about Wiley products visit us at www.wiley.com.

Wiley also publishes its books in a variety of electronic formats and by print-on-demand. Some content that appears in standard print versions of this book may not be available in other formats.

*Library of Congress Cataloging-in-Publication data applied for*

ISBN: 9781118825426 [hardback]

Cover Design: Wiley

Cover Image: © Aleksandra Drizo

Set in 10/12pt Minion by SPi Global, Pondicherry, India

Printed and bound by CPI Group (UK) Ltd, Croydon, CR0 4YY

10 9 8 7 6 5 4 3 2 1

*Dedicated to my beloved parents Zorče and Aleksandar – my eternal life inspirations, watching from the sky.*

# Contents

# Author Biography

Dr. Drizo has 25 years of experience in research, development, implementation, and assessment of innovative technologies for sustainable water treatment and management. She pioneered passive filtration systems for phosphorus removal and harvesting from wastewater, and has led and managed projects in Europe, Canada, New Zealand, USA, Taiwan, and Brazil. She was an invited speaker for over 100 lectures and presentations speaking about solutions for eutrophication mitigation and control and other global environmental challenges.

Dr. Drizo held a number of academic positions across continents. She was an associate professor in sustainable water management at the University of Vermont, USA (2004–2012), a visiting professor at the Massey University, New Zealand (2004/2005), the Ecole des Mines de Nantes, France (2010) and National Pingtung University of Science and Technology (NPUST), Taiwan (2010). She was also a lecturer on Water/Energy Nexus and Global Water Issues at Singularity University, USA (2012 and 2014) and more recently, she worked as a full Professor in Water Technology at Heriot-Watt University, UK (2014–2017).

# Acknowledgements

Finding solutions to protect water resources from phosphorus pollution has occupied me throughout most of my professional life. I sincerely hope that this book will serve and inspire the next generation of phosphorus and water-protection practitioners and regulators.

Foremost, I would like to express my sincere gratitude to my publisher, Wiley-Blackwell, and their incredible team.

I am grateful most of all to Dr. Ian Francis, former Senior Commissioning Editor of the Wiley-Blackwell New Analytical Methods in Earth and Environmental Sciences Series, for acknowledging the importance of the phosphorus pollution problem and for asking me to write this book.

Unforeseeably, from the beginning of writing the book, my life was also filled with the deep sorrows of losing my mother, my father, and my dearest friend and at times it was very difficult to find the strength to continue the work on it. In addition, I had to change jobs and move across continents.

For completing the work in spite of these difficult years, I wish above all to thank Ms. Sonali Melwani, Team Lead and Project Editor of Wiley-Blackwell , for her patience, her constant encouragement, and kind understanding of my tardiness.

I would also like to thank the Production Editor, Ms. Priya Subbrayal and her team, Ms. Carol Thomas, Freelance Copyeditor, for her expertise and time in polishing the manuscript, and Mr. Andrew Harrison, Senior Commissioning Editor at Wiley-Blackwell, for their help and assistance with the final stages of the book.

I owe warm thanks to Eamon Twohig, Program Manager at the Agency of Natural Resources, Department of Environmental Conservation, Vermont, USA, for the proofreading of Chapter 2 and for providing valuable input. He was also 'my right hand' and at the core of our research team during my time at the University of Vermont.

I would like to use this opportunity to thank my colleagues for reviewing and providing sources and permissions for the material described in the book: to Professor Gunno Renman, KTH Royal Institute of Technology, for his review of the Polonite-based systems section and to Professor Tore Krogstad, Norwegian University of Life Sciences, for his review of the Filtralite P section; to Professor Paul Withers, Lancaster Environment Centre at Lancaster University, UK, for his permission to use the map presented in Figure 3.2.; to Mr. Karl-Gustav Niska, Skandinavisk Ecotech AB, for providing pictures of Polonite phosphorus traps (Figure 3.3); to Professor Peter Jenssen, Norwegian University of Life Sciences, for the permission to use his photo of the Filtralite P system (Figure 3.4) and to Mr. Eric Fehlhaber, Conservation Manager at the Sheboygan County Planning & Conservation, Wisconsin, USA, for the permission to use the photos presented in Figure 3.6.

Last but not least, I would like to thank my sister Maja for always being there for me. And to Hugo, for years of happiness, for sharing my dreams and for being by my side.

# List of Abbreviations

| | |
|---|---|
| **AFO** | animal feeding operations |
| **AMPs** | agricultural management practices |
| **APWA** | American Public Works Association |
| **BAT** | best available technology |
| **BCT** | best conventional pollutant control technology |
| **BEMP** | Best environmental management practice |
| **BMAA** | beta-N-methylamino-L-alanine |
| **BMPs** | best management practices |
| **BOD** | biochemical oxygen demand |
| **CAFO** | confined animal feeding operations |
| **CAP** | the Common Agricultural Policy |
| **CNMP** | Comprehensive Nutrient Management Plan |
| **CSA** | critical source areas |
| **CSOs** | combined sewer overflows |
| **CSS** | combined sewer system |
| **CW** | Constructed Wetlands |
| **CWA** | US Clean Water Act |
| **EC** | Electric coagulation |
| **EEA** | the European Environment Agency |
| **EIP** | European Innovation Partnership |
| **ELG** | the Effluent Limitations Guidelines and Standards |
| **EMAS** | Eco-Management and Audit Scheme |
| **EPEC** | the European Public-Private Partnerships Expertise Centre |
| **ESPP** | the European Sustainable Phosphorus Platform |
| **ETV** | environmental technology verification |
| **EQOs** | Environmental Quality Objectives |
| **EU** | European Union |
| **EWWM** | European Wastewater Management |
| **FAO** | Food and Agriculture Organisation of the United Nations |
| **FHWA** | Federal Highway Administration |
| **FTTA** | the Federal Technology Transfer Act, USA |
| **GAPs** | good agricultural practices |
| **GAEC** | good agricultural and environmental condition practices |
| **GEF** | Global Environment Facility |
| **GPRI** | Global Phosphorus Research Initiative |

| | |
|---|---|
| **HABs** | harmful algae blooms |
| **HABaH** | harmful algae blooms and hypoxia |
| **HABHRCA** | the Harmful Algal Bloom and Hypoxia Research and Control Act |
| **HRT** | hydraulic residence times |
| **IFAS-EBPR** | Integrated Fixed-Film Activated Sludge Systems with Enhanced Biological P Removal |
| **ISBD** | the International Stormwater BMPs database |
| **ISO** | International Organization for Standardization |
| **JRC** | Joint Research Centre |
| **LCPB** | Lake Champlain Basin Program |
| **LDH** | layered double hydroxides |
| **LID** | low impact development |
| **LWA** | light-weight aggregates |
| **MBRs** | membrane bioreactors |
| **MUCT process** | Modified University of Cape Town process |
| **MWP** | Ministry of Water and Power, China |
| **MWWTE** | municipal wastewater treatment effluents |
| **MWWTF** | municipal wastewater treatment facilities |
| **MWWTP** | municipal wastewater treatment plants |
| **NJCAT** | New Jersey Corporation for Advanced Technology |
| **NOAA** | the National Oceanic and Atmospheric Administration |
| **NPDES** | the National Pollution Discharge Elimination System in the USA |
| **NPS** | nonpoint pollution sources |
| **NRR** | nutrient recovery and reuse |
| **NFS** | the National Sanitation Foundation, USA |
| **NWRP** | National Water Resources Policy, Brazil |
| **NWRMS** | National Water Resources Management System, Brazil |
| **OECD** | the Organization for Economic Cooperation and Development |
| **OSS** | onsite septic systems |
| **P** | Phosphorus |
| **PE** | Population Equivalent |
| **PEDs** | Performance Enhancing Devices |
| **POWTs** | publicly owned sewage treatment works |
| **PPP** | the Polluter-Pays Principle |
| **PRC** | P retention capacity |
| **PRTT** | Phosphorus Recovery Transition Tool |
| **RBMP** | River Basin Management Plans |
| **ROI** | return on investment |
| **SEI** | Stockholm Environment Institute |
| **SEPA** | Scottish Environment Protection Agency |
| **SORA** | the State Onsite Regulators Alliance |
| **SSA** | steel slag aggregates |
| **SuDS** | sustainable drainage systems |

| | |
|---|---|
| **SWAT** | Soil and Water Assessment Tool |
| **TAPE** | the Washington State Department of Ecology's Technology Assessment Protocol |
| **TMDL** | Total Maximum Daily Load |
| **TN** | total Nitrogen |
| **TP** | total Phosphorus |
| **TSS** | total suspended solids |
| **USA, US** | the United States of America |
| **US EPA** | the US Environmental Protection Agency |
| **USDA** | US Department of Agriculture |
| **USDA NRCS** | the US Department of Agriculture Natural Resources Conservation Service |
| **UWWD** | Urban Waste Water Directive |
| **VBS** | Vegetative Buffer Strips |
| **WWTPs** | wastewater treatment plants |
| **WEF** | the Water Environment Federation |
| **WERF** | the Water Environment Research Federation |
| **WFD** | the EU Water Framework Directive |
| **WIPO** | World Intellectual Property Organization |
| **WIPs** | Watershed Implementations Plans |

# 1

# The Looming Threat of Eutrophication

## 1.1 Introduction

The word 'eutrophic' originates from a word *eutrophy*, from Greek *eutrophia* meaning nutrition and *eutrophos* which means well-fed. Eutrophication has many different definitions depending on whether they describe solely the process of nutrient enrichment or whether they also include impacts and problems caused by such enrichment. In its simplest form eutrophication is defined as the over-enrichment of receiving waters with mineral nutrients, phosphorus, and nitrogen. It results in excessive production and growth of autotrophs, in particular algae, cyanobacteria (Box 1.1), and aquatic macrophytes (Correll 1998; Ansari et al. 2011; van Ginkel 2011). The increased bacterial populations and vegetation abundance result in high respiration rates leading to hypoxia (oxygen depletion). Hypoxia and algal blooms (Figure 1.1) are the two most acute symptoms of eutrophication (Ansari et al. 2011; UNEP 2017).

Hypoxia or oxygen depletion in a water body often leads to 'dead zones' – regions where levels of oxygen in the water are reduced to a point that can no longer support living aquatic organisms (Figure 1.1). Hypoxia in the northern Gulf of Mexico is defined as a concentration of dissolved oxygen less than 2 mg/l (2 ppm). In other oceans of the world, the upper limit for hypoxia may be as high as 3–5 mg/l. The new knowledge on oxygen depletion (hypoxia) and related phenomena in aquatic systems has been recently reviewed by Friedrich et al. (2014).

*Phosphorus Pollution Control - Policies and Strategies*, First Edition. Aleksandra Drizo.
© 2020 John Wiley & Sons Ltd. Published 2020 by John Wiley & Sons Ltd.

Box 1.1 Cyanobacteria

*Cyanobacteria*, commonly referred to as 'blue-green algae' are microorganisms that structurally resemble bacteria. However, unlike other bacteria, they contain chlorophyll a and are the only photosynthetic prokaryotes able to produce oxygen. They are the oldest oxygenic phototrophs on Earth and include nearly 2000 species. Cyanobacterial blooms are highly visible, widespread indicators of eutrophication. Many of cyanobacteria species produce cyanotoxins that are toxic to humans and animals. The most common algae toxins found are microcystins associated with *Microcystis, Anabaena, Oscillatgoria, Nostoc, Hapalosiphon,* and *Anabaenopsis* species (Whitton and Potts 2000; National Toxicology Program 2017). Microcystis is one of the most common bloom formers in freshwater systems on every continent except Antarctica. This genus can produce a suite of potentially harmful compounds including toxins anatoxin-(a) and beta-methylamino-L-alanine (BMAA) (O'Neil et al. 2012).

(a)    (b)

(c)

**Figure 1.1** Examples of harmful algae blooms (HABs) in: (a) Lake Champlain in Philipsburg, Quebec, Canada–USA border; (b) Missisquoi River, a tributary of Lake Champlain, Vermont, USA; and (c) Tubarao Lagoon, Vitoria, Brazil. Source: A. Drizo.

## *1.1.1 Natural versus Cultural (Anthropogenic) Eutrophication*

Eutrophication can be caused by both natural and anthropogenic processes and activities and it is very important to distinguish between the two. *Natural eutrophication* is a process caused by the incursion of nutrients from natural sources. It is a slow, inevitable, and irreversible process, where accumulation of nutrients in the water and the bottom sediments occurs gradually, over hundreds or thousands of years. Over time, it may result in an ultimate transformation of an aquatic ecosystem into a terrestrial biome (van Ginkel 2011; Ghosh and Mondal 2012). *Cultural eutrophication* (Box 1.2) is caused by anthropogenic activities – human, social, and economic, and their interactions.

---

Box 1.2 Cultural (Anthropogenic) Eutrophication

*Cultural eutrophication* is caused by human activities and their perpetual addition of nutrients to the environment. Unlike natural eutrophication, it is often a rapid process which can take place over several years or decades. Cultural eutrophication has been recognized as the single greatest cause of water quality deterioration in freshwater and coastal marine ecosystems worldwide (Smith and Schindler 2009; Schindler 2012). Although it has taken only 60 years for humans to create cultural eutrophication in many freshwater systems, some studies suggest their recovery may take 1000 years under the best of circumstances (Carpenter and Lathrop 2008).

---

The largest and well-established cause of cultural eutrophication is population growth and its incessant demand and pressure on land and water resources. Some of the examples include (i) changes in settlement patterns in urban and rural areas and subsequent land use; (ii) augmented food production and supplies, with the subsequent increase in fertilizer applications for crop growth and confined livestock and inland and marine aquatic farm production operations; (iii) increased wastewater volumes, which exceed existing wastewater treatment infrastructure capacities often resulting in illicit discharges; (iv) increased construction of infrastructure for water storage and transportation, including catchment and/or watershed alterations, such as building dams; (v) increased demand for leisure activities such are golf courses with high fertilizer demand; and (vi) increased use of fertilizer for home gardens and lawns.

Cultural eutrophication is considered controllable, because humans can take measures to reduce and/or minimize the negative impacts of their activities on the environment (van Ginkel 2011). Therefore, this type of eutrophication is the major topic of this book.

## 1.2 Trophic Classes of Water Bodies

Since the beginning of twentieth century, water bodies have been categorized according to their *trophic status*, based on concentration of chlorophyll, the transparency of water, and phosphorus mean concentration

Table 1.1 OECD criteria for trophic status of lakes.

| Trophic status | Total P (µg/l) Mean | Chlorophyll a (µg/l) Mean | Max | Transparency (m) Mean | Min |
|---|---|---|---|---|---|
| Ultra-oligotrophic | <4 | <1.0 | <2.5 | >12 | >6 |
| Oligotrophic | <10 | <2.5 | <8 | >6 | >3 |
| Mesotrophic | 10–35 | 2.5–8 | 8–25 | 6–3 | 3–1.5 |
| Eutrophic | 35–100 | 8–25 | 25–75 | 3–1.5 | 1.5–0.7 |
| Hypertrophic | >100 | >25 | >75 | <1.5 | <0.7 |

Source: OECD (1982). Reproduced with permission of OECD.

(Foundation for Water Research 2006). For standing waters, the Organisation for Economic Cooperation and Development (OECD), proposed a classification scheme consisting of three main trophic classes: oligotrophic for nutrient poor waters, mesotrophic for waters slightly to moderately enriched with nutrients, and eutrophic for waters excessively enriched with nutrients (OECD 1982). In addition they also proposed two boundary classes, ultra-oligotrophic class for the extreme nutrient deficiency and hypertrophic for extreme eutrophication (Table 1.1).

However, the classification for flowing waters according to their trophic status has been debated for the past few decades (Foundation for Water Research 2006). For example, Dodds et al. (1998) reviewed different classification systems for streams and proposed a simplified method based on frequency distribution of nutrients and chlorophyll to define three trophic categories.

The US EPA (2000a) proposed a two-phase approach where streams are initially classified according to their physical parameters associated with regional and site specific characteristics, including climate, geology, substrate features, slope, canopy cover, water retention time, discharge and flow continuity, channel morphology, and system size.

The following phase involves further classifying by nutrient gradient. However, their trophic state classification focuses primarily on chemical and biological parameters including concentrations of nutrients, algal biomass as chlorophyll a, and turbidity, and may also include land use and other human disturbance parameters (US EPA 2000a). More recently, the European Council Directive 91/271/EEC concerning urban waste-water treatment stated that if a river's soluble Phosphorus (P) concentrations exceed 100 µg/l it is excessively enriched with phosphorus (European Commission 2016c).

## 1.3 The Role of Phosphorus in Eutrophication

Phosphorus (P) is an essential component of nucleic acids and as such it has a central role in nearly all biochemical functions of every living organism (Correll 1998; Wyant et al. 2013). P is vital for plants for the synthesis of genetic material, phospholipid membranes, and metabolism (Plaxton and Lambers 2015). In contrast to carbon and nitrogen, P has no gaseous phase; it is found in the atmosphere only in small particles of dust. Phosphorus is present in the environment as a variety of organic and inorganic compounds. In the aquatic systems it can be found in three forms, as the free *ortho-phosphate* ion ($HPO_4^{2-}$), as a *polymer* of phosphate compounds ('*polyphosphates*') or as part of an *organic* phosphorus molecule such as DNA (nucleic acid). However, of the three forms only ortho-phosphate is bioavailable, e.g. it can cross the algal cell membranes and be assimilated by bacteria, algae, and plants (Correll 1998; Wyant et al. 2013).

In aquatic systems P is considered a 'growth-limiting' factor (Box 1.3) because it is usually present in very low concentrations (Schindler 1977). Typical soluble P concentrations in unspoiled rivers are often bellow 30 µg/l. Therefore, the additions of very small quantities of P (0.01 to 0.02 mg P/l) are sufficient to induce harmful algae blooms (HABs) in surface waters (Heathwaite and Dils 2000).

The role of phosphorus in causing eutrophication was first demonstrated by Schindler and his research team, over four decades ago (Schindler 1974). In order to elucidate the primary cause of eutrophication, they conducted a series of whole-lake experiments in the Experimental Lakes Area of north western Ontario, Canada. The lake was divided into two basins, of which one was fertilized with phosphorus, nitrogen, and carbon, and the other one only with nitrogen and carbon. Whilst the former one was covered by algal bloom within two months, in the later one there were no changes in species or algae quantity providing clear evidence that phosphorus has a vital role in eutrophication (Schindler 1974).

Schindler and co-researchers continued the whole ecosystem experiment in which they investigated roles of carbon, nitrogen, and phosphorus in controlling eutrophication for nearly four decades. The results from this long-term research corroborated their previous findings that the only way to reduce eutrophication is to decrease inputs of P (Schindler et al. 2008). However, despite evidence, scientists continued to debate the contribution of nitrogen and carbon to eutrophication, particularly in estuaries and coastal environments (e.g. Howarth and Marino 2006).

---

Box 1.3 The Law of the Minimum

The Law of the minimum concept was developed Carl Sprengel in early 1800s and later promoted by Justus von Liebig (van der Ploeg et al. 1999). *The law represents the origin of the theory of mineral nutrition of plants* and suggests that plant growth is controlled, not by the total amount of nutrients or resources available, but by the availability of the scarcest resource (limiting factor). Over the past century ecologists developed the concept for an aquatic system stating that plant and bacterial growth in an aquatic system would ultimately become limited by the availability of an essential element. This element would then constitute the limiting nutrient for that system at that time, and inputs of that nutrient could be managed to minimize and limit eutrophication (Correll 1998).

More recently, in the light of new European Union Water Framework legislation to control both nutrients (Chapter 3), Schindler (2012) revisited the controversy of which nutrient is the major cause of eutrophication. He discussed the common misconceptions and errors that affected scientific researchers' recommendations including (i) the assumptions that results from the short-term experiments where nutrients are added to small bottles or mesocosms can be extrapolated to whole ecosystems over long time periods; (ii) the investigations about strategies to reverse eutrophication are often made by adding instead of decreasing nutrients from water; and (iii) flawed logic and assumptions about ecosystem-scale nutrient cycling. Furthermore, he pointed out that despite claims that reducing nitrogen is essential to decreasing estuarine eutrophication, there is no documented case history of where this measure has been successful. Regarding criticisms on the lack of response to P control measures in some estuaries, he underlined the well-known fact that the largest human populations and the most intensive conversions of catchments have occurred in coastal areas. Therefore, in these catchments it may take decades to see the response to phosphorus reduction interventions and eutrified waters return to original, phosphorus-limited state (Schindler 2012).

A recent report by the Lake Winnipeg Basin Initiative and published by Environment and Climate Change Canada (ECCC 2017) provided further evidence that nitrogen reduction measures have no or minimal impact on reducing P levels in lakes. The report revealed that after spending $18 million on nitrogen reduction measures between 2012 and 2017, the amount of phosphorus entering the lake fell by less than 1% (ECCC 2017).

### 1.3.1   Phosphorus Pollution Sources

The annual discharge of P to the environment from human excreta (i.e. faeces and urine), greywater (bathroom and kitchen sinks, showers, baths, washing machines) and livestock animals globally is 44.8–45.1 million metric tons. In addition, pets (dogs and cats) contribute around 7–10 million metric tons annually, just in North America and EU countries, resulting in and annual discharge of 52–55 million metric tons of P (Box 1.4).

This discharge is dispersed via numerous sources which are broadly classified as:

- *Point* (end of a pipe) discharges and
- *Nonpoint* (*diffuse*) pollution sources

Both point and nonpoint discharges may originate from a variety of sources, including: municipal wastewater treatment facilities (largely sewage consisting of human wastes), agricultural (animal wastes, pesticides and fertilizers, agricultural surface and subsurface runoff from animal production areas such are farmyards, feedlots and composting piles, and crop fields), onsite residential septic systems (containing human wastes, detergents, other organic wastes from food; septic systems drainage [leachate]

Box 1.4  Phosphorus Discharges into Environment

- A single adult in the industrialized world excretes 1–3 g P/day (as phosphate).
- About 3–3.3 million metric tons of P is generated annually in human excreta (i.e. faeces and urine) and greywater (bathroom and kitchen sinks, showers, baths, washing machines), globally (Mihelcic et al. 2011).
- Taking into account that a single cow excretes 35–50 g P/day and a pig 55 g P/day (Wyant et al. 2013), livestock animals generate an additional 41.8 million metric tons of P annually resulting in a total of 44.8–45.1 million metric tons of P.
- In North America and EU pets (dogs and cats) excrete additional 7–10 million metric tons of P annually, particularly in urban areas. An average dog produces about 340 g of waste per day containing 0.25% of P as $P_2O_5$. In 2017 about 182 million dogs and 207 million cats lived in homes as pets in North America and the EU (Statista 2018a).
- Phosphorus (P) concentrations measured in the wastewater streams and runoff (Figure 1.2) exceed quantities that induce HABs 100–60 000 fold, discharging vast quantities of P into the environment daily and further increasing eutrophication risks.

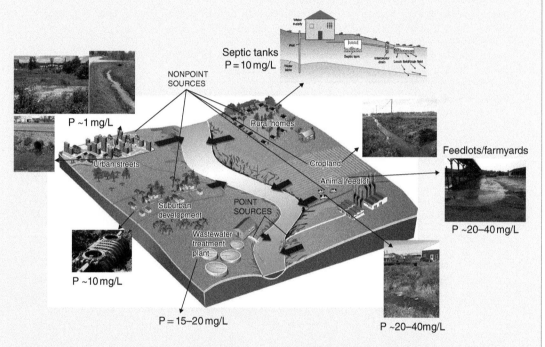

**Figure 1.2** Typical P concentrations measured by the author and co-researchers in various point and nonpoint pollution streams across the USA. Source: A. Drizo. All photo images were taken by the author.

fields), industrial (chemical, organic, and thermal wastes), and urban and suburban runoff from car parks, commercial buildings, and houses (roofs and gardens), construction sites, golf-courses, and roads (Figure 1.2).

## 1.4  Impacts of Eutrophication

Eutrophication has many detrimental impacts on the environment, health (animal and human), and the economy (Table 1.2).

Table 1.2 Impacts of eutrophication.

| Impact | References |
|--------|-----------|
| *Intensified growth and production of algae, cyanobacteria (blue-green algae) and aquatic plants* typically appearing as algal scums or floating mats of plants and commonly referred to as 'algal blooms'. This excessive abundance in vegetation and bacteria increases respiration rates causing significant fluctuations in dissolved oxygen concentrations and water transparency, eventually leading to hypoxia | e.g. Correll (1998), Smith and Schindler (2009), Ansari et al. (2011), Ghosh and Mondal (2012) |
| *Fish deaths and reduced biodiversity.* Low dissolved oxygen causes loss of invertebrates and fish and through their decay, algae and bacteria proliferation, further reducing oxygen content of water and loss of biodiversity | Correll (1998), Ronka et al. (2005), Hautier et al. (2009), Ansari et al. (2011), Brownlie (2014) |
| *Toxins excretion.* Certain algal species in marine and freshwaters including cyanobacteria produce toxins that may seriously affect the health of fish, birds, and mammals. This can occur either through the food chain, or direct contact, or ingestion of the algae. Recent studies revealed that most cyanobacteria produce the neurotoxin beta-N-methylamino-L-alanine (BMAA) which had been linked with the development of neurodegenerative diseases (Alzheimer's and Parkinson's diseases, and Amyotrophic Lateral Sclerosis [ALS]) | Briand et al. (2003), Foundation for Water Research (2006), Banack et al. (2010), Brand et al. (2010) |
| *Aesthetics.* Eutrophication causes increased turbidity, unpleasant odours, slimes and foam formation, diminishing the aesthetic value of waters | Ansari et al. (2011) and Ghosh and Mondal (2012) |
| *Considerable economic losses.* Algal blooms reduce potable water supplies, property values, tourism, and recreation. The losses of local economies due to eutrophication were estimated at $2.2 billion per year in the USA in 2009, and between £75 and £114.3 million per year for England and Wales in 2003 | Dodds et al. (2009), Pretty et al. (2003), Brownlie (2014) |
| *Global Climate Change* will promote cyanobacterial growth and exacerbate algal blooms at much larger scales, further diminishing water availability and potable water supplies | Pearl and Huisman (2008), Moss et al. (2011), Paerl and Paul (2012), Jacobson et al. (2017) |

## 1.5 The Extent of Eutrophication

The occurrence of cyanobacterial harmful algal blooms was first reported in an Australian lake 140 years ago (Francis 1878). During the past century and a half, the eutrophication of fresh and saline waters has been in continuous expansion across the globe (Vollenweider 1970; Schindler 1974; Selman et al. 2008; UNEP 2017; World Resources Institute 2018).

According to the World Resources Institute researchers the number of coastal areas worldwide experiencing symptoms of eutrophication and/or hypoxia increased 85% in just five years between 2008 and 2013 – from 415 to 762. Of the 762 sites, 479 were identified as experiencing hypoxia, whilst

228 sites showed other symptoms of eutrophication, including algal blooms, species loss, and negative impacts to coral reef congregations. The remaining 55 sites were systems in recovery from hypoxia (World Resources Institute 2018). It is estimated that as much as 78% of the assessed continental US coastal area and approximately 65% of Europe's Atlantic coast exhibit symptoms of eutrophication (Figure 1.3). The actual magnitude is most likely much greater given that in many regions of the world (e.g. Asia, Latin America, Africa) research and evidence collection on eutrophication is relatively recent (Le et al. 2010; Walton 2010; van Ginkel 2011; Kundu et al. 2015; World Resources Institute 2018).

In Europe, the European Environment Agency (EEA) declared eutrophication a pan-European problem of a major concern more than 20 years ago (EEA 1995). Despite all the efforts and vast investments, it remains a major threat to achieving the good status of waters required by the Water Framework Directive (European Commission 2015a). The bottom of the Baltic Sea represents the world's largest hypoxic 'dead zone'. Carstensen et al. (2014) recently reported a 10-fold increase of hypoxia in the Baltic Sea in the past 115 years, from 5000 to about 60 000 km$^2$ which they attributed mainly to increased inputs of nutrients from land.

The second largest zone is located in the USA, in the northern Gulf of Mexico, just under the Midwestern Corn Belt. Over the past 30 years (1985–2014) it had an average size of about 13 650 km$^2$ (or 5300 mile$^2$). However, in summer of 2017 the area of this dead zone increased 1.7 fold, to 22 730 km$^2$ or 8776 mile$^2$ (Charles 2017).

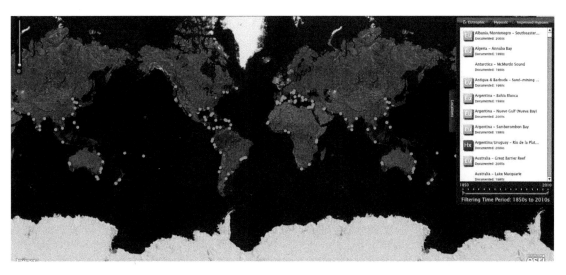

**Figure 1.3** Interactive map of eutrophication and hypoxia. Source: WRI (2018).

## 1.6    Global Climate Change and Eutrophication

There has been increasing scientific evidence that rising nutrient inputs and global warming mutually intensify eutrophication symptoms (e.g. Pearl and Huisman 2008; Moss et al. 2011; O'Neil et al. 2012; Paerl and Paul 2012; Jacobson et al. 2017). This has been attributed to the fact that higher temperatures stimulate cyanobacteria growth and proliferation even at lower nutrient concentrations. For example, increase in surface water temperatures strengthens the vertical stratification of lakes, reducing vertical mixing and promoting accretion of dense surface cyanobacteria blooms. In turn, the blooms may intensify absorption of light, further increasing water temperatures (Pearl and Huisman 2008).

Global climate change also affects patterns of hydrological cycles, e.g. precipitation and drought. These changes could further enhance cyanobacterial proliferation and dominance. For example, more intense precipitation may increase surface and groundwater nutrient discharge into water bodies. Although the freshwater discharge may prevent blooms by flushing in a short term, they will intensify during periods of droughts (longer water residence times). In temperate climates and northern hemisphere this scenario typically occurs when elevated winter–spring rainfall and flushing events are followed by periods of summer droughts (Pearl and Huisman 2008). Recent scientific evidence has shown that eutrophication has also substantially altered fish assemblages and aquatic food webs around the world (Moss et al. 2011; Jacobson et al. 2017).

Despite extensive research during the past four to five decades, many key questions in eutrophication science remain unanswered. Much is yet to be understood concerning the interactions that can occur between nutrients and ecosystem stability. Recent research suggests that nutrients may strongly influence the fate and effects of other non-nutrient contaminants, including pathogens and emerging contaminants (Subirats et al. 2018). There is also limited evidence that eutrophication may promote climate change via greater release of methane from deoxygenated waters and sediments (e.g. Moss et al. 2011).

## Further Reading/Resources

DUJS (2012). Eutrophication in the Gulf of Mexico: how Midwestern farming practices are creating a 'Dead Zone'. Posted 11 March 2012. https://sites.dartmouth.edu/dujs/2012/03/11/eutrophication-in-the-gulf-of-mexico-how-midwestern-farming-practices-are-creating-a-dead-zone/ (accessed 10 April 2019).

Fogg, G.E., Stewart, W.D.P., Fay, P. et al. (1973). *The Blue-Green Algae*. London: Academic Press.

# 2

# Water Quality Legislation and Policy for Phosphorus Pollution Control

## 2.1 Introduction

Policy and regulation play a critical role in empowering technological development and innovation process – from preliminary research to technology diffusion on the market (Kemp 2001; Ashford and Hall 2011; OECD 2018a). The major environmental protection policy, the Polluter-Pays Principle (PPP) was introduced nearly a century ago (Box 2.1). Its' global implementation would bring considerable benefits in protecting water quality from pollution and illegal discharges.

Policies, and in particular those dealing with environmental protection, are always formulated with a view towards the available technology options for dealing with the particular problem. Therefore, as national environmental policies are influenced by the currently available technologies, technology availability plays a vital role in the policy development process in all stages from formulation to the implementation and development of compliance options (Kemp 2001; Ashford and Hall 2011; OECD 2018a).

However, when attempting to solve environmental pollution problems, the fact that national policies are based on currently available technologies creates complex issues and colossal obstacles to novel technologies development and market introduction. On one hand, it has been recognized that innovative products, processes, and services specifically developed to solve environmental pollution problems may reduce the emission of pollutants and protect natural resources, human health, and biodiversity. They can also foster both economic and social development (e.g. OECD 2011, 2013, 2018a, 2018b). Yet, the fact is that the potential inclusion of any novel technology into national strategies would require the revision and adaptation of

*Phosphorus Pollution Control - Policies and Strategies*, First Edition. Aleksandra Drizo.
© 2020 John Wiley & Sons Ltd. Published 2020 by John Wiley & Sons Ltd.

Box 2.1  The Polluters-Pay Principle

The PPP states that those who produce pollution should bear the costs of managing it to prevent damage to human health or the environment (OECD 1992; Munir 2013). Nearly 50 years ago (1975), at Community level, the polluter was defined as the person who directly or indirectly causes deterioration of the environment or establishes conditions leading to its deterioration.

In 1972, the Organization for Economic Cooperation and Development (OECD) recommended the PPP as the 'Guiding Principle Concerning the International Economic Aspects of Environmental Policies'. In 1973 the Council of the European Communities approved the First Program of Action on the Environment and the PPP was made one of the principles of Community's environmental policy. The PPP became globally accepted as one of the principles of environmental policy in many countries and regional and international conventions (OECD 1992; Munir 2013). However, In the USA, the PPP has not been fully implemented nor has it been recognized as a distinct principle or a policy (Lockhart 1997).

priorities, needs, and instruments, which takes a very long period of time (OECD 2011, 2018a). This is best illustrated in the example of the development of policies to support micro Combined Heat and Power in Germany which took 30 years (OECD 2011).

Without a regulatory requirement to deal with a particular environmental pollution problem, the potential recognition, consideration, or inclusion of innovative technologies capable of solving these problems (e.g. Climate Change Mitigation; Water Pollution Prevention, or Reduction) into national and international policies and strategies as 'available technologies' is extremely complex and time consuming. Moreover, even in a rare instances when the inventor succeeds in obtaining his or her technology approved by the accredited environmental technology verification (ETV) authority, its implementation remains hindered for years and/or decades due to absence of national strategies and regulatory requirements for technologies to deal with the particular type of pollution. For example, in the past 10+ years several technologies and products have been developed in the USA that demonstrated capability to reduce P pollution from agricultural and residential onsite septic systems effluents; however, their deployment on the market has been hindered by the extremely long and arduous process intersected with numerous financial and regulatory obstacles and the absence of regulatory requirements. Such a situation results in continuous water pollution from all point and nonpoint pollution sources (NPS). It also prevents the growth of small businesses offering products for water pollution prevention, and consequently diminishes job creation, economic growth, and development.

## 2.2   Water Policies to Protect Water Quality from Phosphorus Pollution

Current Water Policies to protect water quality from pollution are insufficient and inadequate, globally (UNEP 2016; UN Water 2017). Some of the established challenges and barriers to achieving water quality improvements include: (i) poor data quality and quantity (both data collection sites, density, and frequency); (ii) inadequate or absence of national strategies and practices for pollution reduction; and (iii) socioeconomic and policy constraints (e.g. UNEP 2016; UN Water 2017; Voulvoulis et al. 2017). Moreover, even when considerable national and international strategies and commitments are present for a long period of time, such is the case with the EU Water Framework Directive 2000/60/EC (WFD), there are continuous and persistent problems with their implementation. As a result, although considered the most substantial and ambitious piece of European environmental legislation to date, the WFD has failed to deliver its main goal of reducing pollution (including phosphorus) and improving water quality (Voulvoulis et al. 2017).

Responses to water pollution can be broadly divided to political, legal, economic, and social (Kraemer et al. 2001). One of the common political approaches concerning water pollution abatement is the adoption of Environmental Quality Objectives (EQOs). The EQOs define target values for key ambient quality parameters and are subsequently used to evaluate existing environmental conditions.

Until the 1990s the USA and Europe used a uniform standards approach, which served to set limits on a common coordinated basis to deal with water pollution problems. This approach led to significant pollution reduction from point pollution sources, in particular for municipal and industrial wastewater effluents. However given the costs of implementation of the pollution reduction measures, in early 2000 both the USA and the European Union (EU) changed their pollution reduction approaches from the prevention at the source to in-situ ambient water quality (Boyd 2000; Kraemer et al. 2001). This change in the approach had very negative implications on water pollution reduction (including P) and in achieving water quality improvement goals (Christian-Smith et al. 2012; UNEP 2016; Voulvoulis et al. 2017).

### 2.2.1   Water Policies for P Pollution Control – the United States of America

The basis of the US Clean Water Act (CWA) was the Federal Water Pollution Control Act, enacted in 1948. The Act was significantly reorganized and expanded in 1972 and became commonly known as the CWA (US EPA 2018d). In the same year, the National Pollution Discharge Elimination

System (NPDES) was created (Section 402 of the CWA) and became the central piece of US water quality regulations.

The NPDES programme requires polluters to obtain permits, or licences to discharge; the permits specify pollution amounts that can be legally discharged. Depending on the type of the pollutant, discharge limits are determined based on the 'best conventional' control technology or 'best economically achievable' technology (Boyd 2003; US EPA 2018d). However, there is no reference to innovative 'alternative' technologies. Moreover, the NPDES programme only addresses pollution from point sources leaving NPS completely neglected. In addition, NPS became recognized as the primary cause of the water quality impairment (US GS 1999; Boyd 2000).

In the past 20 years there have been several major rules and a few legislations and amendments enacted in an attempt to deal with the ever-increasing phosphorus pollution and subsequent harmful algae blooms and hypoxia (HABaH) occurrences (Table 2.1).

In the past two decades since the establishment of the *Harmful Algal Bloom and Hypoxia Research and Control Act* (HABHRCA), there have been a series of amendments and considerable investments to the NOAA and other agencies (US Department of Agriculture (USDA), US EPA, the National Science Foundation [NSF]) to administer the HABaH programme and combat eutrophication. The NOAA alone received in average $20 million per annum over the past 20 years for the program administration. The US Government Accountability Office reported that just in a 2 year period (2013–2015) 12 US Federal Agencies expended approximately $101 million to 'various HABs activities' (US GAO 2016). The USDA agency awarded $1.8 billion between 2009 and 2016 for use of 45 practices intended to prevent fertilizer runoff. Of these, agricultural producers in Texas, Kansas, Oklahoma, Indiana, and Nebraska were pledged the most funding. The highest amount of funding was allocated to the farmers in Sussex County, Delaware, a top chicken-producing area, received $17 million over the seven years, the most of any US county (Flesher 2017). At the State level, the State of Vermont received the highest federal funding support of $45 million to improve water quality of Lake Champlain (USDA 2014).

Despite considerable financial investment and governmental agencies efforts, not only that there has been no improvement in mitigating nutrient pollution and subsequent HABaH, but there has been recurrent and increased occurrence across the country. For example, between 2014 and 2016, there have been several major HABs incidences and subsequent damages to public health and economy, including the algal toxins incursion to public drinking water supply in the city of Toledo, Ohio (2014) and Ingleside, Texas in 2016 (US EPA 2016b). Moreover, in the past several years HABs occurrences have spread in all 50 states from east coast to west coast (Olson-Sawyer 2017; US EPA 2018b). Yet, regulatory requirements to reduce P pollution have currently been developed only for large municipal and

**Table 2.1** US Water policy rules and regulations related to P pollution and eutrophication issues.

| Date | Rule/legislation | Source |
|------|------------------|--------|
| 1996 | US EPA proposed new rules for implementation of the Total Maximum Daily Load (TMDL) programme in the NPDES for point sources. TMDL 'specifies the amount of a particular pollutant that may be present in a waterbody, allocates allowable loads among sources, and provides the basis for attaining or maintaining water quality standards'. | Boyd (2000) |
| 1998 | Congress authorized and embedded as the Public Law the *Harmful Algal Bloom and Hypoxia Research and Control Act* (HABHRCA) administered by the National Oceanic and Atmospheric Administration (NOAA). Appropriations of $15 million (1999), $18.25 million (2000) and $19 million (2001) respectively for NOAA in support of the comprehensive effort to prevent, reduce, and control HABaH. | US Government (1998) and NCCOS (2018). |
| 2001 | US EPA developed and published water quality criteria for nutrients according to section 304(a) of the CWA. This was a starting point for states and authorized tribes to develop nutrient criteria for their own jurisdictions. | US EPA (2001) |
| 2004 | *The Harmful Algal Bloom and Hypoxia Research and Control Amendments Act of 2004* (Section 104e, Public Law 108–456) reaffirmed and expanded the mandate for NOAA to advance the scientific understanding and ability to detect, monitor, assess, and predict harmful algal blooms and hypoxia (HABaH) events. | NCCOS (2018) |
| 2009 | Development of the criteria was extended to lakes and reservoirs across for each of the eco-regions the country. | US EPA (2018b,c) |
| 2014 | *The Harmful Algal Bloom and Hypoxia Research and Control Amendments Act of 2014* modified the HABHRCA of 1998 and authorized the appropriation of $20.5 million annually through 2018 for the NOAA to mitigate the HABaH. It also called for the creation of the national HABaH programme and 'a comprehensive research plan and action strategy' to combat HABaH, working with state, local, tribal, and international governments for both coastal and inland waters. The action strategy identified the 'specific activities' that the programme should carry out, which activities each agency in the Task Force would be responsible for, and the parts of the country where even more specific research and activities addressing HABaH would be necessary. | US Government (2014) and NCCOS (2018) |
| 2015 | Congress amended the Safe Drinking Water Act and directed the EPA to develop a strategic plan for assessing and managing risks associated with algal toxins in drinking water provided by public water systems. | US Government (2015) |
| 2017 | The most important amendments made in the *The Harmful Algal Bloom and Hypoxia Research and Control Amendments Act of 2017* were to: <br>• Add Army Corps of Engineers to the list of agencies on the Task Force. <br>• Combine the sections on freshwater and coastal algal blooms, and requirement that scientific assessments be submitted to Congress every five years for both types of water <br>• Establish a website that would provide information about the HABaH program activities to 'local and regional stakeholders' <br>• Require the Task Force to work with extension programmes to promote the programme and 'improve public understanding' about HABaH <br>• Require the use of 'cost effective methods' when carrying out the law <br>• Require the development of 'contingency plans for the long-term monitoring of hypoxia'. <br>• Fund the programme and the comprehensive research plan and action strategy from 2019 through 2023. <br>In addition a new section was added which allows federal officials to 'determine whether a HABaH event is an event of national significance'. | US Government (2017) |

industrial wastewater treatment plants (WWTPs). They do not exist for P loading from other sources (agricultural effluents and runoff, onsite residential septic systems, urban storm water runoff).

Instead of the development of water quality standards for each pollution source (based on known average P concentrations as shown in Figure 1.2) and an appropriate regulatory framework that would require and enforce P reduction from each of these sources (according to the PPP, Box 2.1) the US EPA continues to promote Total Maximum Daily Loads (TMDLs) and the ambient Numeric nutrient criteria as a critical tool for protecting and restoring a water body's designated uses (US EPA 2018b).

In fact, the introduction of TMDLs in 1996 (Table 2.1) brought a number of negative effects on the US water quality. The most detrimental one has been the fact that TMDLs initiated the major shift in the water quality regulations and changed the entire approach of attaining pollution reduction from the technology-based, end-of-pipe treatment to reduce pollution (at least from the point sources) to an 'ambient' water monitoring and standards. Therefore, instead of working on the development of limit values to prevent pollution from known sources (e.g. small scale wastewater treatment facilities, residential septic systems, animal farm operations, agricultural tile drains, urban/rural storm water runoff, Figure 1.2) water quality standards development has been focused on the receiving water bodies only (e.g. rivers, lakes, reservoirs segments, and areas) and in some cases for large wastewater treatment facilities. The ambient approach to water standards, combined with the absence of regulatory requirements, and reluctance to adopt PPP do not oblige nor motivate any of the residents in a given watershed area (home owners, farmers, businesses) to reduce the P pollution they create and discharge in the very water body that is supposed to reach/comply with certain P values ('water quality standards') set by the TMDL.

Unless the regulators make a fundamental change in the current ambient water quality approach and make effort to develop values/permissible levels for P concentrations at the origin of the pollution source any decrease in algae blooms occurrence or other improvement in water quality will be highly unlikely. This can be very well illustrated with an example of Lake Champlain, Vermont presented in Case Study 2.1.

---

Case Study 2.1  Lake Champlain, Vermont, USA

Lake Champlain is the sixth largest body of freshwater in the USA and has been severely impaired by phosphorus pollution and HABs for over four decades, in particular in the northern part of the Lake (Missisquoi and St Albans Bays) (Figure 1.1; Figure 2.1 and Table 2.2). The lake is shared between states of Vermont and New York and internationally with Canada (Quebec).

Following the suit of Conservation Law Foundation (CLF) against the EPA in 2008, the EPA was required to revise and develop a new TMDL and nutrient water quality standards to protect water quality of Lake Champlain (Table 2.2). During the 10 years long TMDL negotiating process (2008–2018), one of the key milestones has been a shift from 'best' or 'accepted' agricultural practices (BMPs/AAPs) to required agricultural practices (RAPs) with a Rule promulgated in December 2016 which is fundamental to mitigating P pollution from agricultural lands. However, the Vermont Agency of Agriculture, Food, and Markets (VAAFM) with limited resources, has established inspection schedules for medium and small farm operations every three and seven years, respectively. Instead the proposed rules to date focus on flood prevention and control and are insufficient to protect water quality from P pollution (Weber 2018).

**Figure 2.1** Examples of harmful algae blooms in Lake Champlain at the mouth of Rock River, Interstate 89. Source: A. Drizo, summer 2013.

**Table 2.2** History of the Lake Champlain phosphorus TMDL.

| Year | Event/Proposed Regulation/Rule | Source* |
|---|---|---|
| 1990 | Numeric total P concentration criteria proposed for segments of Lake Champlain in the Vermont (VT) Water Quality Standards | Smeltzer (2013) |
| 1993 | VT Water Quality Standards adopted and endorsed by New York (NY) and Quebec as mutual lake management goals in a signed Water Quality Agreement. | Smeltzer (2013) |
| 1996 | • Completion of a joint VT and NY P budget and modelling study for Lake Champlain<br>• Negotiation of a P reduction agreement for Lake Champlain between VT and NY, including preliminary point- and nonpoint-source load allocations by watershed. | Smeltzer (2013) |
| 2002 | *EPA approved joint Vermont and New York Lake Champlain Phosphorus TMDL.* The TMDL established an annual average load limit, calculated at an effluent P concentration of 0.6 mg/l for all facilities in Vermont that discharge at least 200 000 gal of wastewater per day. The aerated lagoon facilities over this size limit were required to meet a P effluent concentration of 0.8 mg/l. | Smeltzer (2013) and Winsten (2004) |
| 2004 | Initiation of the *VT Clean and Clear Action Plan* to fund the implementation of the Lake Champlain TMDL and address similar water quality needs statewide. | Smeltzer (2013) |
| 10/2008 | Conservation Law Foundation (CLF) filed suit against EPA seeking to set aside the approval of the 2002 VT TMDL claiming it is inconsistent with the federal Clean Water Act (CWA) | Conservation Law Foundation, CLF (2018) |
| 2010 | VT ANR Completion of a Revised Implementation Plan for the Lake Champlain Phosphorus TMDL, as required by Act 130 (2008). | Smeltzer (2013) |
| 01/2011 | EPA withdrew its approval of Vermont TMDL | CLF (2018) |
| 2011–2015 | EPA and the State of Vermont worked on development of a new TMDL | US EPA (2016b) and CLF (2018) |
| 5/2012 | VT Legislature passed Act 138, which transfers certain rulemaking authority from the Water Resources Panel of the Vermont Natural Resources Board to the Agency of Natural Resources (VT ANR). | VT ANR (2018) |
| 5/2014 | State releases VT Lake Champlain P TMDL Phase 1 Implementation Plan | CLF (2018) |
| 10/2014 | VT ANR released Vermont Water Quality Standards (Environmental Protection Rule Chapter 29(a)) | VT ANR (2018) |
| 8/15 | EPA releases the final draught P TMDL for VT Segments of Lake Champlain | CLF (2018) |
| 2015 | *Vermont Legislature passes VT's CWA (Act No. 64 [H.35])* | VT ANR (2018) |
| 6/2016 | EPA, working with VT DEC, issued new TMDL for 12 Vermont Lake segments to address P loading | US EPA (2016a) |
| 1/2017 | VT ANR released updated Vermont Water Quality Standards | VT ANR (2018) |
| 3/2018 | Agency of Agriculture files proposed subsurface tile drainage amendment to required agricultural practices (RAPs) rule | VT Digger (2018) |
| 5/4/2018 | EPA issued VT Report Card on Water Quality focusing on progress made | VT ANR (2018) |
| 24/4/2018 | CLF issues their Report Card on VT Water Quality failed progress | Weber (2018) |

Source: Modified from Smeltzer (2013).

Not only that during the past decade VAFFM and VT USDA Natural Resources Conservation Service (NRCS) did not respect the obligation to develop the RAPs for P reduction (RAPs, also known as 'best management practices' and GAPs, Chapter 3), but despite vast funding available (USDA 2014; Vermont EPSCoR 2018) they also failed to implement and assess the performance of the innovative passive filtration system which had been accepted as an interim conservation practice in 2013. The practice is known as *P removal system interim code #782* and has been invented with the aim to intercept subsurface (tile) flow, ground water or surface runoff flow, and reduce the concentration of phosphorus (Drizo 2010, 2011b, 2011c and Section 3.5.13).

Following previous investment of $46 million, in 2014, the USDA NRCS provided an additional $46 million to VT farmers as a 'new five-year pledge to address continuing water quality issues' via conservation activities on and around farming operations in the Missisquoi Bay, St. Albans Bay, and South Lake Watersheds (USDA 2014). The University of Vermont received $40 million ($20 million in 2011 and another $20 million in 2016) from the NSF (Vermont EPSCoR 2018).

Despite these investments of over $100 million, there had been no reduction in P loading to the Lake nor HABs outbreaks nor any improvement in Lake Champlain water quality. And yet, VT State continues to receive substantial investments from federal budget and other sources aiming to solve P pollution and algae HABs problems. For example, in autumn 2017 Lake Champlain Basin Program (LCPB) in conjunction with the New England Interstate Water Pollution Control Commission (NEIWPCC) announced a call for grant proposals offering considerable amounts of funding for a variety of pilot projects to improve Lake Champlain water quality. These included, but were not limited to $300 000 per project for technical solutions and 'innovative strategies' to reduce nutrient loading to Lake Champlain. In addition, $500 000 per project was offered for projects 'ranging from innovative planning or farm management programs to installation of management practices demonstrated to be successful elsewhere that have not yet been widely explored or implemented in the Lake Champlain Basin (LCBP 2017)'. However, the project duration set in the request for proposal was only nine months. Moreover, project start date was May 2018 and end date March 2019, meaning that at the best any management practice that would be installed would not be tested for its performance during the time of the greatest runoff and P loading, e.g. spring snowmelt. Therefore the Funding agency offered $500 000 for projects that would not even last long enough to evaluate their performance in nutrient reduction during all four seasons.

Given such approach it is not surprising that despite considerable financial investment, HABs outbreaks continue to intensify in northern portion of Lake Champlain, as well as in other lakes in the Area. Lake Carmi, for example, located in the town of Franklin, in the northwest corner of VT was declared 'Lake in crisis' in February 2018 (Bill H.730) due to severe HABs and hypoxia (Polhamus 2018; Vermont Government 2018). The Bill authorizes VT ANR to declare a lake to be in crisis when the condition of the water poses a potential harm to the public health, poses a risk of damage to the environment or natural resources, or is likely to cause significant devaluation of property value or other significant negative economic effects. Since February 2018, ANR became a designated State agency with sole authority for regulation of water quality requirements, whilst the State shall appropriate financial resources to ANR sufficient to perform the work (Vermont Government 2018).

### 2.2.1.1 Agricultural Policy for Phosphorus Pollution Prevention and Control – USA

The US EPA's major efforts and roles in reducing nutrient pollution can be view on their website (US EPA 2018b). The activities related to Agricultural P pollution mitigation and their practical applications are summarized in Table 2.3.

Following the report on the poor state of national waters and extent of nutrients pollution the US EPA Office of Water issued a memo calling for the establishment of the 'State(s) Nutrients Framework as a tool to guide ongoing collaboration between EPA Regions and states in their joint effort to make progress on reducing nitrogen (and P) pollution' (e.g. US EPA 2011,

**Table 2.3** The USA EPA's major activities in agricultural pollution prevention.

1 *Working with states to identify water bodies impaired by nitrogen and phosphorus pollution and to develop TMDLs to restore or protect waters.*

As discussed in Section 2.2.1, development of TMDLs has serious negative implications on water quality. Some of the major ones are that it does not prevent and address pollution at the source and the time period (in some cases decades) it takes to develop. The most serious one is that none of the polluters (e.g. neither farmers nor citizens) are required to implement any conservation practice to reduce P loading from their lands. The implementation is optional and voluntary.

2 *Administering a wastewater permit programme that establishes discharge limits and monitoring requirements necessary to protect water quality standards and the environment from point sources of nutrient-related pollutants – i.e. from municipal and industrial facilities, concentrated animal feeding operations (CAFOs), and stormwater.*

The National Pollution Discharge Elimination System (NPDES) programme and creation of water quality standards and TMDLs is discussed in Section 2.2.1. According to the EPA NPDES CAFO are considered as a point source, including any discharge of pollutants from land spread manure (US EPA 2006). In 1999, the US Department of Agriculture (USDA) and the EPA issued a joined Unified National Strategy for Animal Feeding Operations (USDA 1999). The Strategy required that all animal feeding operations (AFO) owners and operators must establish a site specific Comprehensive Nutrient Management Plan (CNMP). In 2002 the EPA proposed a new regulation, the Effluent Limitations Guidelines and Standards (ELG) for feedlots (beef, dairy, swine, and poultry), which aimed to set an allowable limit for discharges of manure, wastewater and other process waters under the existing NPDES CAFO permit process (US EPA 2003a). Requirements from the ELG were incorporated into a CAFOs NPDES permit to set standards for the volumes of discharge allowed for specific pollutants under the permit (US EPA 2003b). However, the implementation of practices and technologies to meet ELGs remains on voluntary basis.

3 *Providing states, the regulated community, and the public with guidance on the regulatory requirements of nutrient management plans for regulatory requirements of nutrient management plans for CAFOs.*

These regulations were designed to ensure that AFO implemented technologically sound and economically sustainable nutrient management plans (NMP's) that would protect water resources from excess P loading from farms. Nutrient management planning involves accounting for and recording all the nutrients on the farm, determining what nutrients farmer will need, and planning how, how much, when and where to apply them to the crop land. Its major focus is to provide advice to farmers that may prevent over-application of nutrients thus protecting water quality and minimizing the impact on the environment whilst still provide optimum yield for economic benefit. *It is important to note that whilst NMPs may contribute to minimizing of agricultural P pollution from crop lands, it does not require farmers to implement any conservation practices to reduce P pollution from any of the sources (farmyard and feedlots runoff, silage leachate, manure piles, surface and subsurface runoff).*

2013). They highlighted that their major focus in promoting pollution reduction is on promoting various conservation practices, commonly referred to as 'good agricultural practices (GAPs)' or 'BMPs' (US EPA 2011). In addition, the EPA underlined their efforts in promoting innovative approaches to accelerate the implementation of agricultural practices, including so called targeted stewardship incentives, certainty agreements for producers that adopt a suite of practices, and nutrient credit trading markets. The conservation practices and their implementation are discussed in detail in Chapter 3, Section 3.5. Agricultural Phosphorus Pollution and Mitigation Measures and Strategies.

In 2016, The US EPA published a 'Renewed Call to Action to Reduce Nutrient Pollution and Support for Incremental Actions to Protect Water Quality and Public Health', which highlighted the fact that Nutrient pollution remains one of the greatest challenges to US water quality and presents a growing threat to public health and local economies (US EPA 2016b).

Although the EPA underlined their efforts in promoting innovative approaches to accelerate implementation of agricultural practices in both 2011 and 2016 memos, there are numerous obstacles to both the innovation and in particular implementation and water quality monitoring of innovative conservation practices. Some of these obstacles as they relate to BMPs for agricultural P pollution mitigation are reviewed in Chapter 3, Section 3.5.15.

### 2.2.1.2 Regulations for Phosphorus Pollution Prevention and Control from Residential Effluents (Onsite Wastewater Treatment and Disposal Systems)

Despite overwhelming evidence in the national and international literature, P pollution from residential septic systems (onsite wastewater disposal and treatment systems) has not been recognized as a sufficiently significant threat to water quality to initiate policy driven regulatory requirements (Chapter 3, Section 3.3). Therefore, despite the fact that 26% of the households in the USA are not connected to the main sewer lines and the ever-increasing reports of HABaH on lakes across the country, there has been no support for research to investigate P loading from onsite septic systems nor for development of potential solutions and technologies that could solve the problem. This approach has greatly reduced economic growth and job creation potential for small businesses offering engineering and environmental consulting in wastewater treatment.

On the other hand, the regulatory framework for onsite treatment systems developed and introduced in Scandinavia since 2006 (Swedish EPA 2006; Eveborn 2013) had considerable benefits for the environment, society, and economy. It initiated research and development of novel P removal technologies at the Royal Institute of Technology (KTH) and several other universities and more than 10 different small businesses (Sweden and Finland) that offer novel products and technologies for P removal and their wide implementation in homes across Sweden to protect inland water resources and the Baltic Sea from P pollution (Chapter 3).

### 2.2.1.3 Regulations for Phosphorus Pollution Prevention and Control from Urban Stormwater Runoff

In 1987 the US Congress broadened the CWA definition of 'point source' pollution to include industrial stormwater discharges and municipal separate storm sewer systems ('MS4') (US GPO 1987). Based on the 1987 Stormwater Amendments, in October 1990 the US EPA issued the stormwater rule, which became effective on 17 December 1990 (55 Fed. Reg.47, 990). The regulation included combined sewer overflows (combined sanitary sewage and industrial treated wastewater mixed with rainwater and land runoff); municipal separate stormwater systems (owned or operated by municipalities that receive only stormwater runoff); separate stormwater systems (those that serve industrial facilities and were historically part of the industry's NPDES permit); and nonpoint source runoff (Franzetti 2016). The US EPA implemented the stormwater program by segregating the classes of stormwater discharges to municipal and industrial. The US EPA's National Stormwater Permit programme, had two phases. Under Phase I, municipal separate storm sewer systems were required to reduce the pollutants in their stormwater discharges 'to the maximum extent practicable' through the use of management practices, control techniques, design and engineering methods, and other appropriate approaches. However, regulations applied only to municipalities serving populations of over 100 000 persons. In addition NPDES permits were required for industrial dischargers and construction sites of 5 acres (20 000 m$^2$) or more. Industrial dischargers are required to implement best available technology (BAT) and best conventional pollutant control technology (BCT) to control their stormwater discharges (Chapter 3). However, nutrient reduction was not required. Instead priority pollutants for BAT included metals and organic compounds. BCT required biochemical oxygen demand (BOD), total suspended solids (TSS), pH and faecal coliform reduction (Franzetti 2016). The Phase II rule was finalized nearly 10 years later (1999) and it expanded requirements for NPDES permits on stormwater discharges from a wider sector of municipalities, industries, and construction sites. In 2003 requirements for NPDES permits were expanded to construction sites 1 acre (4000 m$^2$) or more, and other large property owners (such as school districts). Those smaller than 1 acre were also required to obtain a permit if they were a part of a larger common plan of development or sale disturbing a total of 1 acre of land or greater. Phase II also required municipalities to develop elaborate Stormwater Management Plans (SWMP) and to include six minimum control measures: (i) public education and outreach; (ii) public participation/involvement; (iii) illicit discharge detection and elimination; (iv) construction site runoff; (v) post-construction run-off and (vi) pollution prevention/good housekeeping. Numerous community education programmes were developed and promoted by the US EPA to educate citizens on how to help to minimize the introduction of pollutants into the stormwater system (Franzetti 2016).

In the past 10 years each of the states developed their own standards, rules, and regulations for urban stormwater runoff management (US Office of Water 2011). Whilst most of the states are required to achieve runoff volume and TSS reduction, only some states developed standards and rules for phosphorus reduction: Maine, Vermont, New York, Rhode Island, Pennsylvania, Virginia, and Minnesota. However similar to agricultural runoff management, the implementation of any of the stormwater management practices is voluntary, e.g. there are no regulatory requirements for P reduction from urban stormwater runoff (US EPA 2018c).

## 2.2.2 Water Policy – European Union

Phosphorus legislation in the EU s integrated in several Directives, legislations and action plans, binding for each Member State with the requirement to meet all of the stated objectives (Amery and Schoumans 2014). Amongst several European directives that relate to P issues (Table 2.4), the Urban

Table 2.4 European Legislation related to P pollution issues.

| Bathing Water Directive | 76/160/EEC amended by 2006/7/EC |
| --- | --- |
| Urban Waste Water Directive (UWWD) | 91/271/EEC |
| Nitrates Directive | 91/676/EEC |
| Integrated Pollution Prevention and Control Directive (IPPC) | 96/61/EC |
| Water Framework Directive (WFD) | 2000/60/EC |
| Groundwater Directive (daughter directive of WFD) | 2006/118/EC |
| Marine Strategy Framework Directive | 2008/56/EC |
| Waste Framework Directive | 2008/98/EC |
| Industrial Emissions Directive (replaces IPPC Directive 96/61/EC) | 2010/75/EU |
| Use of phosphates and other phosphorus compounds in consumer laundry detergents and consumer automatic dishwasher detergents | Regulation (EC) No. 648/2004 amended by (EU) No. 259/2012 |

Source: Amery and Schoumans (2014), modified by A. Drizo.

Wastewater Directive 91/271/EEC and the WFD 2000/60/EC have been considered the most important ones.

The WFD has been widely accepted as the most substantial and ambitious piece of European environmental legislation. It consolidated and updated earlier EU water legislation and extended the concepts of river basin management planning to the entire EU. It advocates a combined approach to pollution prevention and control and in this respect integrates with the EU Integrated Pollution Prevention and Control Directive which is the key regulatory initiative controlling emissions from major industrial sectors and the Urban Waste Water Directive (UWWD) which sets minimum standards of treatment for sewerage systems and sewage treatment works (Griffiths 2002).

The major purpose of the WFD was to establish a framework for the protection of European waters in order for Member States to reach 'good status' objectives for all surface and ground water bodies throughout the EU and to prevent any further deterioration. These efforts were based on a six-year cycle, where the WFD environmental objectives were expected to be met by 2015. Member States that did not achieve the objectives had extensions to second (from 2015 to 2021) and third (2021 to 2027) management cycles respectively (Voulvoulis et al. 2017).

In order to reach this ambitious goal, the WFD introduced a major paradigm shift from traditional end-of-pipe solutions towards whole catchment management and systems thinking, which adopts an interdisciplinary, integrated, and holistic approach. It introduced the terms 'catchment-based approach' and 'integrated river basin management', to refer to the management of land and water as a system; and it required EU Member States to manage water at hydrological units, to develop strategic River Basin Management Plans (RBMP) and operational Programmes of Measures (PoM) and, whilst doing so, to engage with stakeholders and the wider public (e.g. Boeuf et al. 2016; Voulvoulis et al. 2017).

Despite its ambitious plans, nearly two decades since it was adopted, the WFD has not delivered its objectives. Some of the major causes have been recently reviewed and thoroughly discussed by Boeuf et al. (2016) and Voulvoulis et al. (2017). Moreover, despite continuing deterioration of the water quality caused by increasing occurrence of algae blooms and hypoxia P removal is required only from large municipal sewage treatment plants (European Commission 2016c). There have been no amendments to UWWD which requires the collection and secondary treatment of wastewater only in agglomerations greater than 2000 population equivalents (PE, Box 2.2.) and more advanced treatment for agglomerations >10 000 PE (>2000 m³/day) in designated sensitive areas and their catchments. As a result, of the 540 major cities in Europe, only 15% have advanced tertiary sewage treatment (e.g. processes to remove nutrients), whilst 31.1% do not have any form of wastewater treatment (Visvanathan 2015).

---

**Box 2.2** Population Equivalent

In wastewater treatment, population equivalent (PE) or unit per capita loading, is the number that expresses the ratio of the sum of the pollution load produced during 24 hours by industrial facilities and services to the individual pollution load in household sewage produced by one person in the same time (European Commission 2016c). In Europe, it is the term most used to design and describe package sewage treatment plants. 1 Population Equivalent or 1 PE represents 200 l of flow containing 60 g of biochemical oxygen demand (BOD).

---

Moreover as advanced treatment (e.g. P removal) is required only for agglomerations >10 000 PE it excludes a vast majority of other major P pollution sources (Section 1.3.1) as well as sewage treatment works for all agglomerations discharging 1999 m³/d and below which may discharge as much as 40 kg P/d. Given that the addition of just 10 g of P can induce growth of up to 1 kg of algae (Chapter 1) it is evident that the current P pollution regulations set by the UWWD have serious limitations in protecting European surface and ground waters from HABs and eutrophication. The Blueprint to Safeguard Europe's Water Resources Report 2012 highlighted the fact that eutrophication remains a major threat to achieving the WFD good status of waters (European Commission 2015a).

In order to meet the goal for 'good status of waters' set by the WFD, the possibility of requiring new and more stringent consent for P discharges (e.g. 0.1 mg/l large wastewater treatment facilities and 0.5 or 1 mg/l for smaller facilities) has been considered in the EU Member States (e.g. Pidou 2015; Brockett 2016). This initiated investigations of the novel technologies and/or modifications of currently available technologies for their ability to meet proposed stringent P limits by a number of corporations specializing in wastewater treatment products. The key water treatment companies including Thames Water Treatment Utilities Ltd., Severn Trent Water, Veolia Water Technologies,

Suez Advanced Solutions UK, BluewaterBio Ltd., RWB Special Products RV presented their P removal products at the tenth European Wastewater Management Conference held in Manchester, UK in October 2016.

However, even if these new products and technologies would be able to meet the requirements of the newly proposed stringent P regulations for sewage treatment works these products would only be implemented at agglomerations over 10 000 PE (generating 2000 m³/d wastewater). Therefore on the watershed scale this greater P reduction achieved will be negligible compared to total P discharges from all other sources which are excluded from the regulations (e.g. small sewage treatment works, agricultural P pollution, residential, urban, and rural stormwater runoff).

### 2.2.2.1 Agricultural Water Policy – Europe

Similar to the USA, current EU legislation does not adequately address P discharges from agricultural activities. The Common Agricultural Policy (CAP) is centred mainly on water conservation and irrigation (e.g. the investments to conserve water, improve irrigation infrastructures, and enable farmers to improve irrigation techniques). The agricultural water quality protection is regulated only for nitrite and pesticides (European Commission 2017). Amery and Schoumans (2014) recently conducted a thorough review of European Agricultural Policy including CAP, the agricultural subsidies system, and the single farm payment scheme. They underlined that the legislation on P use in agriculture varies widely amongst different countries and regions in Europe. For example, some countries have elaborate regulations which set limitations on P application on agricultural land (e.g. Belgium, Denmark, Estonia, France, Northern Ireland). Other countries or regions either do not have any phosphorus legislation (e.g. England, Wales, Czech Republic) or rely on voluntary measures as part of agri-environmental programmes (Finland), or on the restrictions on the use of manure (Nitrates Directive) to indirectly regulate the use of phosphorus (e.g. Germany, Hungary, Greece, Ireland, Luxembourg).

Similar to the US and Canadian Water Policies, the major strategies recommended by the regulators to minimize P pollution from agricultural lands in Europe are GAPs (Chapter 3, Section 3.5.4) and nutrient management planning (Amery and Schoumans 2014). However, there are large differences in policy construction, implementation, and in the effects from the policy regarding nutrient management legislation amongst European Countries (Jacobsson et al. 2002).

In the past several years there has been an increasing concern over the decline in the global reserves of P (Chapter 4). Due to Europe's limited resources of P and dependence on imports, in 2014, phosphate rock was added to the EU list of Critical Raw Materials, highlighting the need for innovative solutions for P removal, re-use, and recovery from all available waste streams (European Commission 2015b). This initiated interest in novel processes for nutrient recovery from wastewaters, mainly sewage treatment plants and to a much smaller extent highly concentrated agricultural effluents and manure (Chapter 4). Buckwell and Nadeau (2016) recently conducted a comprehensive review of nutrient recovery and reuse (NRR) issues and opportunities in European agriculture. They highlighted the benefits of NRR and the fact that its wide adoption could considerably improve the current nutrient management. However, they concluded the expansion of the NRR concept will take time due to the complexity of the legislation affecting the fertilizer industry, farms, food industry, and the water treatment sector at both the EU and national levels. The European Commission recognizes the need for bioeconomy and circular economy and has been promoting both concepts through its Research and Innovation programmes. Similarly, in 2017 the UN World Water Development Report focused on the potential of wastewater as the untapped resource (Chapter 4). However, the crucial legislation that would bring about much needed change to enable more sustainable water management and NRR has not been developed. Meantime the water quality in Europe and around the world continues to deteriorate.

### 2.2.3   Brazil Water Policy

Brazil is the largest country in South America, encompassing around half of the continent's land area and population (207 million). Brazil's surface water resources account for 50% of the total resources of South America and 11% of the worldwide resources. However, these resources have uneven distribution with some regions having an excess quantities (e.g. the Amazon rainforest), and other regions suffering from water shortages (e.g. the Northeast). Moreover, until recently, the country has had a highly hierarchical and sectorial water management system without any conservation strategy. Therefore despite being endowed with an estimated 11% of the world's fresh water resources Brazil has serious water shortages and water quality issues, in particular in the semi-arid region of the Northeast (Nelson 2008). The first major regulation that contained some management strategy related to natural resources was The Water Code (Código de Águas) of 1934. The code enabled the expansion of Brazil's hydroelectric power system and became popularly known as The 'Water and Mines Code' ('Código de Águas e Minas'). The Code also distinguished private land ownership from the ownership of water and minerals on that land (Nelson 2008).

However, it took over 50 years for Congress to promulgate the National Water Resources Management Act (1988) and enact the Water Resources Management Law, establish National Water Resources Policy (NWRP) and create the National Water Resources Management System (NWRMS) (Nelson 2008; Veiga and Magrini 2013). Similarly to European countries, the NWRP (Law No. 9.433, 1997) introduced an integrated approach with the river basin as the territorial unit for the implementation of the NWRP, water as a fragile and finite resource, and decentralized and participatory management of the resource. From a policy perspective, the Law introduced economic instruments, such as the water usage charges. In addition the new system introduced the perception of water as a public good with economic value in order to promote conservation (Nelson 2008; Veiga and Magrini 2013). Creation of the NWRMS brought changes at the institutional level as the new water management system developed through different institutions at federal, state, and river basin levels. However, similar to the introduction of TMDLs in the USA and WFD in Europe, two decades since the Law took effect, the implementation process still faces many challenges, hindering the effective consolidation of the instruments set out by the Law (Veiga and Magrini 2013). Although hosting the 2016 Olympic Games pushed the Brazilian government to launch a series of plans to solve severe water quality problems, in particular in Guanabara Bay they failed to achieve them. It has been estimated that 70% of the residents living in the Guanabara Bay watershed still lack basic sewage treatment, especially those living in favelas ('slums'). A number of researchers described eutrophication and hypoxia problems in the bay (e.g. Schwamborn et al. 2004; de Carvalho Aguiar et al. 2011; Fistarol et al. 2015). It has been estimated that the Bay receives 470 t of biological oxygen demand and about 150 t of industrial wastewater daily, which originates from 17 000 different industries including pharmaceutical and refineries, oil and gas terminals, and two ports. In addition it receives 18 t/day of petroleum hydrocarbons and 10 000 t/month of hazardous substances (Fistarol et al. 2015). Von Sperling (2016) recently conducted a comprehensive review of urban wastewater treatment in Brazil. He presented a survey of treatment systems performed by the National Water Agency including sewage collection and treatment coverage, number of WWTPs according to process type, and their distribution of according to process and population size. In addition he reviewed the most widely used treatment processes in Latin America. Unlike in Europe and USA, the largest number of WWTPs in Brazil are for small towns; for example of the 2187 plants surveyed, 25% were serving populations lower than 2000 inhabitants, almost 50% populations up to 5000 inhabitants, and 80% were for populations less than 20 000 inhabitants. However, none of the treatment plants are equipped to provide nutrient removal (von Sperling 2016).

As in the USA and Europe, P pollution from agricultural wastewaters and runoff and urban stormwater runoff is not regulated. And like in China, Brazil had a rapid expansion of livestock between 1993 and 2003. During this decade, there was 33% increase in the number of housed animals (i.e. beef, dairy cows, swine, and poultry), most in the South Region of Brazil (i.e. Paraná, Rio Grande do Sul, and Santa Catarina States) where 43% and 49% of Brazil's swine and poultry production is located, respectively. Brazil is the second biggest exporter of cattle beef in the world, with colossal quantities of manure deposited over grazed pastures. In addition, there has been a 4.5-fold increase in fertilizer P use on agricultural lands in the past 40 years. Sharpley (2016) recently reviewed the management of P fertilizer applications to agricultural land in Brazil, and compared two different agricultural practices for sustainable land management: no-till and conventional tillage. The implementation of policies for environmental protection from agricultural activities is under jurisdiction of the Ministry of Agriculture, Livestock, and Supply (Ministério da Agricultura, Pecuária e Abastecimento, MAPA). However, similar to other countries, P pollution from agricultural land is not regulated.

### 2.2.4   China Water Policy

China's water policy has evolved several times in the past 30 years. The beginning of Chinese water policy occurred in 1978 which was the year of China's economic reforms. In 1984, the State Council established the Ministry of Water and Power (MWP). The MWP released their first national water resources assessment in 1987 (Xu 2016). The main Laws and Regulations in China's Water Policy were summarized by Lee (2004) (Table 2.5).

The rapid economic growth of China from 1999 to 2009, resulted in eightfold increase in GDP per capita. The rapid expansion of industry, agriculture, and urbanization also resulted in severe environmental pollution (Box 2.3).

In 2005 livestock production become a pillar in China's agricultural and rural economy and a top priority target for continuing rapid development and modernization in governmental five-year plans. Therefore, China become the world's largest livestock producer with more than 400 million cattle, sheep, and goats producing an estimated 4.8 billion tons of manure annually (Kaufmann 2015). In 2011 the annual dairy sales in China exceeded US$4 billion and were estimated to continue to increase by 30% per year.

Table 2.5 Selected laws and regulations in China's water policy.

| Title | Year |
| --- | --- |
| Environmental Protection Law | 1989 |
| Solid Waste Environment Pollution Prevention & Control Law | 1995 |
| Water and Soil Conservation Law | 1991 |
| Water Pollution Prevention Law | 1984, revised 1986 |
| Water Pollution Discharge Permit Management Measures | 1988 |
| Drinking Water Protection Area Pollution Prevention Management Rules | 1989 |
| Water Law | 1998, revised 2002 |
| Water Pollution Prevention Law Implementation Measures | 2000 |
| Environmental Impact Assessment Law | 2002 |
| Cleaner Production Law | 2002 |

Source: Modified from Lee (2004).

Box 2.3  China Pollution Facts

- Approximately one-third of industrial wastewater and over 90% of household sewage in China was released into rivers and lakes without any treatment. Nearly 80% of China's cities (278) had no sewage treatment facilities and underground water supplies were contaminated in 90% of the cities.
- In summer 2011, the China government reported 43% of state-monitored rivers were unsuitable for human contact due to severe pollution. One study found that 8 of 10 Chinese coastal cities discharge excessive amounts of sewage and pollutants into the sea, often near coastal resorts and sea farming areas (Facts and Details 2014).
- A study by China's Environmental Protection Agency conducted in February 2010 reported that water pollution levels were double what the government predicted them to be mainly because agricultural waste was ignored. The first pollution census conducted in 2010 revealed that farm fertilizer was a bigger source of water contamination than factory effluent (Facts and Details 2014).
- In 2011, half of China's population does not have access to safe drinking water. Nearly two-thirds of China's rural population, i.e. over 500 million people, use water contaminated by human and industrial waste containing dangerous levels of arsenic, fluorine, and sulphates.
- In 2014, it was estimated that 980 million of China's 1.3 billion people drink partly polluted water. More than 600 million Chinese drink water contaminated with human or animal wastes whilst 20 million people drink well water contaminated with high levels of radiation. A large number of sites with arsenic-tainted water had been discovered. China's high rates of liver, stomach, and oesophageal cancer have been linked to water pollution (Facts and Details 2014).

However, poor farming practices and a large population of livestock animals, became one of the major causes of the severe environmental pollution and health problems. The results from the First National Census on Pollution Sources in 2009 showed that agriculture represented a bigger source of pollution in China than industry. Researchers reported that farming was responsible for 44% of chemical oxygen demand (organic compounds in water), 67% of phosphorus, and 57% of nitrogen discharges into bodies of water (Kaufmann 2015).

Recognizing the magnitude of environmental pollution and degradation, and scarce water resources, since 2000 the Chinese government has invested billions of dollars to combat environmental problems. In 2014, the government announced plans to invest $330 billion to fight water pollution (Reuters 2014). In April 2015, the State Council issued a most comprehensive 'Water Pollution Prevention and Control Action Plan' (known as the 'Water Ten Plan') which was the result of coordination and inputs from more than 12 different ministries and government departments (Han et al. 2016). The plan aimed to greatly reduce the pollution of severely polluted water bodies, improve the quality of drinking water, and reduce the over extraction of groundwater. In total, there are 238 specific actions involved, however, the plan set 10 general measures and identified responsible government departments for each action as well as the deadlines. The plan covered four broad actions: (i) Control pollution discharge, promote economic and industrial transformation, and save and recycle resources; (ii) Promote science and technology progress, use market mechanisms, and enforce law and regulations; (iii) Strengthen management and ensure water environment safety; and (iv) Clarify responsibilities and encourage public participation. It also proposed a change from the previous approaches to China's water crisis that focused on large-scale engineering solutions for the provision of clean water (e.g. Han et al. 2016).

However, as with the USA and Europe, there have been major challenges in the implementation of the plan. For example, Han et al. (2016) highlighted that whilst data collection and transparency has improved, there remain major gaps that impede accurate assessment of the watersheds' and catchments' surface and groundwater quality. Their comprehensive review and compilation of data on the water quality classes in groundwater and surface water at the regional and national scale revealed considerable water pollution challenges across the country. They also underlined the fact that contamination of deep groundwater is occurring on a much large scale then understood by the authorities (Han et al. 2016). Based on the scientific evidence they proposed several recommendations for improvements in the Water Plan and underlined that water pollution clean-up will take decades.

### 2.2.5   India Water Policy

India is the second most populated country in the world (1.3 billion), but has only 4% of world's renewable water resources covering only 2.6% of the world's land area. Similarly to China, rapid development and urbanization resulted in severe water pollution. Almost 70% of its surface water resources and a growing percentage of its groundwater reserves are contaminated by various biological, toxic, organic, and inorganic pollutants. In many cases, water bodies have been rendered unsafe for human consumption as well as for other activities, such as irrigation and industrial needs. It has been estimated that the absence of, or inadequate, sanitation and hygiene results in the loss of 0.4 million lives annually (Murty and Kumar 2011).

Historically, environmental policy development for pollution prevention and control in India started 50 years ago. The acts that directly addressed water pollution were the Water Act (1974), the Water Cess Act (1977 and 1988), and the Environment (Protection) Act or EPA (1986). The first two Acts represent the foundational legislations in the context of water pollution in the country, whilst the later one (EPA 1986) was designed to fill the gaps for the control of industrial pollution. Today these laws have mainly remained confined to controlling industrial water pollution (Murty and Kumar 2011). Similar to other countries there are no standards and/or regulations to address water pollution originating from agricultural runoff and effluents. Although standards were developed for domestic sewage treatment, in 2011 only 26% of cities and towns have wastewater treatment facilities. Of those only 55% comply with the standards (Murty and Kumar 2011). In 2012 the government issued a new National Water Policy Draft (Singh et al. 2015). The plan included: (i) the imperatives of providing both clean drinking water and adequate resources for irrigation; (ii) the move towards investigating renewable sources of energy like hydro power; and (iii) natural disaster management and rehabilitation following devastating floods and drought. The policy also seeks to offer economic incentives and penalties to reduce water pollution and wastage. An in depth review of the policy is provided by Singh et al. (2015).

## 2.3 Governance of Innovative Technologies for Phosphorus Removal

Due to the lack of information on the governance of innovative phosphorus removal technologies, this section will mainly focus on the general situation regarding innovative water treatment technologies.

Tait and Banda (2016) provided a comprehensive report of the role of regulations, guidelines, and standards in governance of innovative technologies in the UK and EU. Although the focus of this study was on three particular industry sectors (personalized medicine manufacture for autologous cell therapies, industrial biotechnology/synthetic biology, and financial technology, FinTech) the findings of this study are pertinent to innovative technologies in general. The authors investigated the role of guidelines and standards in the implementation of regulations, and how they can support the speedy and effective delivery of innovation that safely meets public needs and desires, and where appropriate, contribute to the development of a thriving innovation environment. Some of the major findings were:

1  the current governance systems for innovative technologies lack coherence and are in need of a new approach to guide more effective decision making;
2  different governance approaches will be required for disruptive and incremental innovation;
3  the relevant elements of a governance approach that need to be addressed will differ across sectors;
4  when a regulatory system is imposed in the early stages of development of an innovative technology, it requires subsequent adaptation but proves difficult to adapt;
5  where guidelines and standards exceed the minimum requirements of a regulation or go beyond reasonable societal expectations this can have serious negative impacts on innovation.

The European Commission's 'Eco-Innovation Action Plan' is described in a comprehensive website which provides information on good practices and policy developments on eco-innovation from around Europe (European Commission 2018d). The website also provides thorough information on the Funding Programmes, Country Profiles (describing actions, achievements, regulatory developments and trends at national level relevant to eco-innovation) and Environmental Technology Verification (ETV) process (European Commission 2018d). One of the most relevant Funding Programs for Water Innovation is Horizon 2020 (Case Study 2.2, Box 2.4).

---

Case Study 2.2  Horizon 2020 Funding Programme for Water Innovation

In 2011, the European Union issued the *Roadmap to a Resource Efficient Europe* (COM [2011] 571) as a flagship initiative for achieving the Europe 2020 strategy. Following issuance of the document, the EU Heads of State and Government called on the European Commission to bring together all of the EU's previous research and innovation funding under a single common strategic framework which led to the establishment of the Horizon 2020 Funding Programme (European Commision 2018f). In December 2013 the very first Horizon 2020 Funding Work Programme (2014–2015) was announced. It had two major Calls aimed specifically to provide support for Water Innovation and Waste Recycling (including nutrients) under Section 12 (Climate action, environment, resource efficiency, and raw materials). Moreover, the major focus of the WATER-1-2014/2015 Call (Box 2.4) was on Water Innovation and Boosting its value for Europe, and in particular, aimed at bridging the gap from innovative water solutions to market replication (e.g. European Commission 2013, 2018f).

The following year (2016–2017) the main objective of the Horizon 2020 Programme Call for Water was to enable bringing innovative water solutions to the market and support the implementation of the objectives of the European

---

Box 2.4  H2020-WATER-2014/2015 Call – Water Innovation: Boosting Its Value for Europe

The challenge highlighted the fact that in 2011, the pollution of water from run-off alone (predominantly of agricultural origin) was estimated to cost the EU EUR 30 billion per annum. It underlined that improvement of the state of water resources, both in terms of quantity and quality, would trigger substantial economic benefits and assist in meeting the objective of the WFD – to achieve good status by 2015. The activities in the call were crafted to address: integrated approaches to water and climate change adaptation and mitigation; bringing innovative water solutions to the market; and harnessing water research and innovation results for the benefit of industry, policy makers, and citizens in Europe and globally. The total available budget was 67 million euro for 2014 and 93 for 2015 (e.g. European Commission 2013, 2018f).

Innovation Partnership (EIP) and the Joint Programming Initiative on Water (European Commission 2018f). However, of the 10 water issues related topics, only one was pertinent to innovative phosphorus technologies: *CIRC-02-2016-2017: Water in the context of the circular economy*. The major requirement was to demonstrate the potential of efficient nutrient recovery from water. Therefore the entire language and terminology used to describe the need and requirement for novel technologies was branded from the word 'pollution and/or phosphorus/nutrients pollution' to 'resource and/or resource recovery and resource efficiency'. Consequently the language used to describe the need for novel technologies shifted from 'nutrients reduction/removal' to 'nutrient recovery' and 'resource efficiency' and 'circular economy'.

The current Horizon 2020 Work Programme (2018–2020) available funding is about €30 billion, making it *the largest single integrated programme of publicly funded research and innovation during this period across the EU*. It ranks comparably with anything being implemented by the EU's counterparts in the world's major centres for research and innovation. The Circular Economy 'Connecting economic and environmental gains' remained one of the four major focus areas, with the estimated budget of €947 million. In addition, the European Innovation Council pilot (EIC pilot) programme has been introduced as a major new component in Horizon 2020 focusing on providing support for innovative firms and entrepreneurs with the potential to scale up their businesses rapidly at the European and global levels. It offers additional €2.7 billion in funding via the SME Instrument, the Fast Track to Innovation (FTI), Future and Emerging Technologies (FET) Open and the Horizon Prizes Opportunities (European Commission 2018f).

As these are the final years of the Horizon 2020 programme, the interim evaluation of the programme was conducted. Whilst it has been recognized that the increased focus on innovation, including the approach to support small and medium enterprises (SMEs) had very positive outcomes, several areas for improvement were suggested. These included: addressing regulatory barriers to innovation, building synergies with other EU instruments and giving special attention to market-creating innovation (European Commission 2018f).

In the USA, the information about technology innovation can be found on the US Environmental Protection Agency (US EPA) websites (US EPA 2018e, 2018f). However, the website describing Technology Innovation for Environmental and Economic Progress (US EPA 2018f) represents 'historical material' which reflects the US EPA website as it existed on 19 January 2017 and is no longer updated.

Although the website describing technology innovation does not provide any note about being discontinued, the site does not function properly (e.g. it shows a display for information on Environmental Technology Innovation Clusters, EPA Technology Transfer and Technology Transfer Roadmap, however only the later one is functional). The EPA Technology Transfer section explains how EPA conducts collaborative research with non-federal partners, protects intellectual property, and licences EPA's technologies through the Federal Technology Transfer Act (FTTA) program. It also provides information on Citizen Science and Crowdsourcing which was developed by the EPA innovation team as Creative Approaches to Environmental Protection. Citizen science covers a suite of innovative tools to engage with the public to apply their curiosity and contribute their talents to science and technology (US EPA 2018f). However, the information on the technology road map is restricted by the EPA. The roadmap is supposed to examine a broad range of approaches to solving the country's most pressing current environmental problems and preventing future ones. The Water Environment Research Federation (WERF), which is the country's leading research organization advancing the science of water, published Technology Roadmap for Sustainable WWTPs nearly 10 years ago (WERF 2010). This document acknowledged technology gaps and highlighted the need to address these gaps systematically by future research efforts. In October 2013, the Water Environment Federation (WEF), Environmental Defense Fund (EDF), and The Johnson Foundation discussed the development of a roadmap for the implementation of a nutrient management vision. They set 'a new goal' for the next generation of wastewater treatment which would have a zero net impact with regard to nutrient discharges by 2040. In 2015, two new documents 'The Nutrient Roadmap Primer' and 'The Nutrient Roadmap Book' were released (WEF 2015a, 2015b). The later document discussed development of the vision for nutrient removal including (i) Regulatory Requirements, (ii) Watershed Water Quality and TMDLs, and (iii) Effects of Nutrient Management. However their only focus was nutrient recovery from WWTPs. In 2013 WEF and WERF jointly established the new initiative titled 'The Leaders Innovation Forum for Technology (LIFT)' with the aim 'to help bring new water technology to the field quickly and efficiently' (Section 2.3).

The eco-innovation agenda in China is fairly recent as they have only recently developed the necessary policies supporting stricter water policies and eco-innovation. Moro et al. (2018) conducted a comprehensive study of the industrial dynamics of water innovation and conducted a comparison between China and Europe. Using patent classifications, they divided water innovations in two distinctive groups: (i) traditional ('general') water solutions mainly related to innovations covering water distribution, water supply, and sewage distribution and treatment and (ii) innovations that are closely related to climate change adaptations and mitigation technologies, e.g. eco-innovative or 'green' technologies. The authors used a regional REGPAT patent data base developed by the OECD which covers over 2000 regions across OECD countries. It contains granted patents filed only at the European Patent Office (EPO), US Patent Office (USPTO) and World Intellectual Property Organization (WIPO). In total 41 699 water patents were extracted covering period 1979–2014.

Data analysis revealed that 'traditional water innovation' (group i) is carried out predominantly by the big companies in both regions. Big companies owned 61.5% and 49% of all water patents granted in Europe and

China, respectively. When it comes to green, eco-innovations, public water innovators (universities) play a much greater role in China than in Europe, with the Chinese universities having 15-fold more patents (31%) compared to European Universities (2%). The authors postulated that some of the reasons for such a large difference can be attributed to the fact that unlike in China, in Europe, water and innovation regulations and policies have been established for several decades; that the existence of regulations facilitated innovation and collaborative partnerships between universities and the private sector and the creation of spin-off companies that became patent owners. To verify their hypothesis, the authors conducted additional analysis of water eco-innovations as a function of the environmental policy index with the aim of verifying the influence of water regulations on the evolution of patenting activities in China and Europe. The results indicated that European patenting activities are in close alignment with the development environmental policy. In China, this alignment is more recent, e.g. patent activities did not follow development of the environmental policy from the beginning. When data were separated to patents filed by Chinese knowledge institutions and those by Chinese companies, the results revealed that the policy development has an overall positive effect on both groups. However, despite the fact the water companies have more patents in total, the policy index showed a greater contribution to an increase in patent activities from knowledge institutions compared to Chinese companies. China's innovation dynamics is constantly changing in all industry sectors. As in many other countries, the Chinese Government has a key impact on how the country is supporting eco-innovation, and is a major driver via the implementation of regulations and investments. Recent positive changes in Chinese policies towards sustainable development to minimize environmental degradation (Section 2.2.4) has shown considerable innovation effect on rapid technological development and catching-up with respect to water technologies.

The information on the governance of innovative water technologies in Brazil is fairly limited. According to UNESCO, although Latin America accounts for 8% of the global population, the subcontinent accounts for only just 1.5% of global business investment in research. They stated that the reason that innovation has played a minor role in the economic development has been the fact that in this subcontinent traditionally the academic sector drives research and development (UNESCO 2017). However, the country is known to have a labyrinth of regulations that are hostile both to innovation and foreign investment (Smith 2015). In the last few years a number of local governments in Brazil started to push for the development of clean tech legislation, including providing land charge discounts for water recycling, waste reduction, or renewable energy use. In 2016, the government introduced new legislation as well as the National Strategy for Science, Technology and Innovation (ENCTI) 2016–2019, with the aim to boost the country's economic performance and increase productivity through innovation. The strategy aimed for gross expenditure on research and development (GERD) to reach 2.0% of GDP in 2019. Detailed information is provided in the STI Outlook Country profile 2016 (Innovationpolicyplatform 2018).

Water in India is governed as a public good, with water business being divided between municipal and government, industrial and commercial, agriculture and household sectors. In 2000, technological options for water management were limited to structural interventions for increasing supplies, whilst technically feasible and economically viable options for cleaning up of aquifers were almost non-existent (Kumar and Ballabh 2000). Khemka (2016) highlighted that although evolving, the awareness of environmental, social, and economic impacts in water governance remains largely disjointed. Moreover, that water-related data availability is fragmented, scattered across multiple agencies, and inadequate for sound decision making. In addition there is lack of information on the interconnectivity of rainwater, surface water, and groundwater, land use, environmental flows, ecosystems, socio-economic parameters, and demographics at the watershed level (Khemka 2016). Thus, the situation does not seem to have changed much despite considerable investments into water innovation from the UK, EU countries, and North America and the establishment of innovative water programmes and partnerships aiming to boost water innovation. For example, in 2012 three Canadian universities and eleven Indian institutions formed a $30 million partnership aiming to improve water and infrastructure safety and eradicate diseases (University of British Columbia 2012). Another example is the Indo-Canadian Science Programme on Clean Water Technology established to promote multidisciplinary research partnerships between the India–Canada Centre for Innovation Multidisciplinary Partnerships to Accelerate Community Transformation and Sustainability (IC-IMPACTS) and various Indian universities and institutions and the Ministry of Science and Technology (IC-IMPACTS 2018).

The UK Research and Innovation India facilitated a 300 fold growth of the UK–India collaborative research portfolio (from £1 million in 2008 to around £300 million) comprising over 140 individual projects, involving over 175 different UK and Indian research institutions and more than 100 industry partners. Following the extensive recent collaboration of a number of EU member countries with India, in the autumn of 2017 the European Commission announced a new Funding Programme EU–India water cooperation. The programme calls for the development of new innovative and affordable solutions for Indian conditions, both in urban and rural areas, to address challenges associated with drinking water purification with a focus on emerging pollutants; waste water treatment, with scope for resource/energy recovery; reuse, recycle and rainwater harvesting, including bioremediation technologies; and real time monitoring and control systems in distribution and treatment systems (European Commission 2018g).

## 2.4 ETV for Innovative Phosphorus Removal Technologies and Practices

The development of ETV programmes originated in North America. ETV started as a voluntary tool in the mid-1990s with the objective to accelerate market acceptance of innovative environmental technologies. As vendor-generated data are often viewed with scepticism, there was a need for a well-founded market-based verification process to provide users with an independent assessment and reliable information about these novel technologies' performance (Merkourakis et al. 2007). In addition to the verification process, the innovators could also apply for their technologies' certification, which would guarantee that specific standards or performance criteria are met (e.g. NSF certifications Section 2.4.1.1).

The USA and Canada developed two very different approaches to ETV: in the US model, the programme managing organization has the testing of the technology entirely performed by one of its partners, a third verification organization. Thus, the US model aims to make public the technology performance data without judging the results. Unlike the US, in the Canadian ETV model, the programme managing organization first collects the claims and data provided by the technology owner and submits them to a third-party verification organization. The organization verifies the data and then compares them with the vendor's claim. Following the USA and Canada, national ETV programmes have spread across the globe (e.g. South Korea, Japan, Europe, China). They follow either the USA or Canadian model (Merkourakis et al. 2007). By 2012, more than 10 active ETV programmes had been established worldwide. As a result, technology owners sometimes had to repeat testing and verification when entering new markets in different countries (Andersson 2012). ETV programmes in the USA and Europe are presented in the following sections.

### 2.4.1   USA

#### 2.4.1.1   Federal Testing Programmes

Established in 1995, the USA ETV programme was abolished in early 2014 (US EPA 2016c). During nearly 20 years of operation it provided credible data on approximately 500 new technologies. Between 1995 and 2004 the US EPA provided over $63 million in funding to ETV. The costs per verification or protocol were nearly $580 000 in 1995 and decreased to ~$120 000 in 1999. In 2007, the ETV moved to a fully vendor/collaborator-paid programme with the US EPA providing only in-kind technical support, quality assurance, programme evaluation, and outreach. Between 2007 and 2010, the USA ETV programme participated as a founding member in the International Working Group (IWG) on ETV working on the development of an international approach to verification that would allow mutual recognition. In 2016, ISO published a new standard ISO 14034:2016, Environmental management – ETV, with the aim of helping companies that are developing innovative environmental technologies reach new markets (Gasiorowski-Denis 2016). However most of the countries still request their own certifications and run their own testing programmes. Moreover, ETV programme testing and verification was only developed for nitrogen and did not provide the possibility for testing any innovative technologies for P removal at the federal level (US EPA 2009; Drizo 2012).

The *National Sanitation Foundation International (NSF International)*, was founded in 1944 from the University of Michigan's School of Public Health as the NSF to standardize sanitation and food safety requirements and protect human health. They are an independent, accredited organization which provides testing, audit, and certification services for products and systems in their testing facilities in the USA, Canada and Europe (NSF International 2018).

They offer a range of standards and products performance certifications for residential and commercial wastewater treatment products. For example, the NSF/ANSI 40 standard established in 1970 sets treatment requirements for Chemical Oxygen Demand ($CBOD_5$), TSS and pH for treatment products treating 400–1500 gpd (1514–5678 l/d). NSF 245 (2007) sets the rules for nitrogen reduction, whilst NSF 350 & 350-1 (2011) was developed for onsite water reuse products. However, similar to the ETV programme, NSF International does not have any standard or testing procedure for P removal products neither for residential nor any other wastewater treatment. In 2010 the costs of NSF certification ranged from $120 000 to $140 000 per application, each lasting 24–36 months.

As the USA has not developed a standard for P treatment requirements, the only way to obtain certifications for innovative P removal products for onsite residential wastewater treatment is to apply for the individual State approvals for use as 'alternative technology' (Drizo 2012). However, these types of approvals only review the literature on the product performance, e.g. they do not include any laboratory testing. Nevertheless obtaining an approval for use as an 'alternative technology' is a complex process that takes 24–30 months on average, even after the person succeeds in identifying the responsible governmental department (Drizo 2012). For example, in Vermont the onsite residential treatment systems and products are approved by the Agency of Natural Resources (ANR), whilst in Virginia by the Department of Health.

In the case of innovative P removal products for agricultural effluents and runoff treatment, the process is even more arduous. The application and approval of the treatment system/product is under the jurisdiction of the US Department of Agriculture Natural Resources Conservation Service (USDA NRCS). The process for innovative P removal technologies approval is illustrated in Case Study 2.3.

Case Study 2.3 Innovative P Removal Technologies and Practices Verification and Approval Process – Example from the State of Vermont, USA

As the evidence on environmental pollution is collected and disseminated via scientific research, in a number of cases, the research on novel technologies for environmental protection is initiated and performed by researchers at Universities. The process of novel technology creation starts with researcher's motivation and idea. The most straight-forward and shortest route representing various stages of research, development and verification is shown in Figure 2.2. Two key assumptions are made: (i) necessary laboratory/field equipment is available at the Research Institution/University; (ii) from the first grant submission onwards, the researcher is always successful in obtaining funding (which is rarely the case). Following the initial idea, the researcher has to obtain sufficient funding and evidence on novel technology performance prior to data and scientific publications presentation to the relevant regulatory agencies. These funding, research, and publication cycles consists of three steps, and each of them has to be completed before the next one can commence: (i) the researcher assembles the team, writes and submits a grant proposal to the relevant funding programme; (ii) response period from the funding agency is usually between six and eight months; (iii) Preliminary laboratory set up building and performance investigations requires at least 6–12 months. Thus the first Cycle (preliminary laboratory investigations period) requires at least two years: Following this period the researcher starts second Cycle. The entire process takes minimum seven years (Figure 2.2).

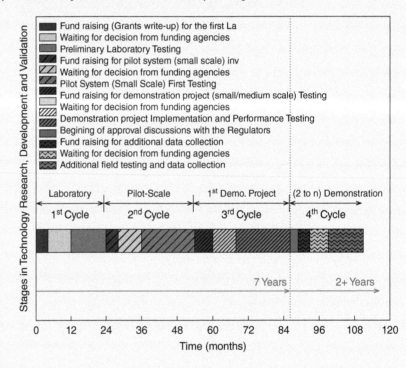

**Figure 2.2** Technology research, development, and validation roadmap.

As if seven years of testing and validation of novel technology is not sufficient, once the third Cycle is finally completed demonstrating technology's beneficial performance in the field at 'appropriate scale' and 'all year round' performance, the regulatory agencies staff will require more data collection (fourth Cycle). For example, they may require testing in different soil type, investigations of potential negative impacts on environment, plans for media disposal (in a case of passive filtration technologies). This period of time is highly variable. It is only upon this fourth Cycle completion that the technology may be considered for the approval process by the appropriate regulatory agency staff. This period may last anywhere from 18 to 36 months (Drizo 2012).

The only way for the researcher/inventor to obtain the right to offer their product/technology for use on farms as a voluntary practice for agricultural P pollution prevention is to obtain status for their solution as the 'conservation practice' from the USDA NRCS. However, once the technology/practice is approved by the USDA NRCS, it becomes 'their practice' with design and implementation being offered by the USDA staff. Thus after years of effort to have their technology recognized as the legitimate conservation practise by the USDA NRCS (Figure 2.2) the inventor is placed in the situation of having to compete with USDA NRCS staff in offering their practice to the very market for which the innovators developed the solution. As the farming community is already familiar with the USDA NRCS staff they will always choose them for a practice implementation over the vendor offering the very same technology they invented.

For example, after nine years of research experience, pilot and demonstration testing across the USA and internationally, and several years of negotiating with the VT NRCS and VAAFM (2010–2013), Drizo's P removal system for surface and subsurface flows has been recognized by the USDA NRCS staff in VT in August 2013. It became the first and only interim conservation practice standard for P removal from agricultural surface and subsurface flows and became known as the USDA NRCS P removal System Code Standard 782 (USDA NRCS VT 2013). In the last few years, P removal System Code 782 (invented and developed by Drizo for the USDA NRCS of VT) received the interim conservation practice standard status in several additional states: Wisconsin (USDA NRCS WI 2015), New York (USDA NRCS NY 2016), Maryland (USDA NRCS MD 2016), and Pennsylvania (USDA NRCS PA 2017). However to date only one single system had been installed in Sheboygan county, WI.

## 2.4.2 Europe

According to the Joint Research Centre (JRC) report, there were no ETV programmes in Europe 10 years ago. Instead, there had been several similar programmes which partially resembled systems for certification, approval, or eco labelling. A comprehensive overview of these programs can be found in the JRC report generated by Merkourakis et al. (2007).

The European Public–Private Partnerships Expertise Centre (EPEC) consortium generated a detailed assessment of the market potential and demand for an EU ETV scheme in June 2011. The final report, the information on the ETV Steering Group, as well as ETV events that took place between 2010 and 2013 can be found at the European Commission archived website (European Commission 2016a).

The EU ETV Pilot Programme became operational in 2013 under the Eco-Innovation Action Plan as a new tool to help innovative environmental technologies reach the market. Similarly to other countries, the EU ETV scheme was initiated in order to facilitate the diffusion of innovative environmental technologies and to advance technology transfer and trade through use of a single instrument at the EU level. Between 2013 and 2016, ETV provided European technology manufacturers access to third-party validation of the performance of their new technologies. At the end of a successful verification process, a Statement of Verification is issued by independent Verification Bodies, summarizing the actual performance of the verified technology and the results of the tests performed. The EU ETV programme thus provides transparent and credible information on the new technology, allowing developers to prove the reliability of their claims whilst giving purchasers a wider range of credible options (European

Commission 2018e). In April 2018 the new EU ETV General Verification Protocol 1.3 was published. The protocol integrates all relevant references to the ISO Standard 14 034, ensuring that an ETV verification in the EU programme is also fully compliant with the ISO Standard for ETV (European Commission 2018d).

# 3

# Phosphorus Removal Methods and Technologies

## 3.1 Introduction

As highlighted in the previous chapter, the regulatory requirements for phosphorous (P) removal have only been developed in North America, the European Union, Australia, New Zealand, and more recently China. Moreover, apart from Scandinavia, they address only municipal wastewater effluents from large cities (e.g. domestic or sanitary sewage consisting of the wastewater carrying faeces and urine, washing water from showers, kitchen and laundry wastes). One of the reasons may be in the facts that (i) municipal wastewater effluents from large cities represent clear, 'end of pipe' point pollution sources, and as such are relatively easy to control; (ii) historically, the focus of wastewater engineering has been on the conveyance and treatment of raw sewage; and (iii) the construction and implementation of municipal wastewater plants (MWWTP) for sewage effluents treatment traditionally has been funded by government funds, e.g. are incorporated into annual income tax and water bills (CCME 2006; European Court of Auditors 2015; Copeland et al. 2016). However, a large number of municipal wastewater treatment plants (MWWTPs) do not provide P removal. For example, Visvanathan (2015) reported that of the MWWTPs implemented in 540 major cities in Europe only 10.5% had processes to remove nutrients. According to the United Nations Environment Programme, about 650 million tons of sewage and 36 000 tons of P are discharged into the Mediterranean Sea annually, of which 70% is untreated (Gunes et al. 2012).

In the USA, there are 14 748 MWWTPs (also called publicly owned sewage treatment works, POWTs) serving 25% of the population (The American Society of Civil Engineers 2017). However, the current number of POWTs with P removal processes is not known and there are no comprehensive

*Phosphorus Pollution Control - Policies and Strategies*, First Edition. Aleksandra Drizo.
© 2020 John Wiley & Sons Ltd. Published 2020 by John Wiley & Sons Ltd.

nationwide data on nutrient discharges and removal at POTWs (US EPA 2018a). The US EPA has just recently started the National Study of Nutrient Removal and Secondary Technologies (US EPA 2018a).

Generally, even when regulatory requirements for P removal exist (e.g. municipal wastewater effluents from large cities), only a fraction of MWWTPs have P treatment processes in place. When it comes to addressing P removal from non-regulated pollution sources (e.g. residential septic systems, agricultural, urban stormwater runoff) the situation is much more severe as it brings about tremendously complex social, economic, and political issues. One of the major barriers to implementation of any onsite treatment systems for P removal (from both point and diffuse pollution sources) is the complete absence of willingness to pay for the technologies/processes to reduce P loading and subsequent Harmful Algae Blooms (HABs)in surface waters. Citizens (homeowners, agricultural producers, small-scale wastewater treatment plant owners, small businesses) expect that the government should fund the installation of products/devices/technologies. However, as governments around the world (apart from Scandinavia) do not recognize P pollution from residential effluents (onsite septic systems, OSS) they do not provide any financial support or incentives to the homeowners for the construction, installation, and maintenance of P removal products and processes from this pollution source. Similarly, unless recognized as 'the good agricultural practice' (Section 3.5.) the implementation of any product to remove P from wastewaters and runoff originating from agricultural activities would have to be financed by the agricultural producers (at least in part, as there are certain government programmes that offer to fund through grants, Section 3.5). The situation is particularly complex when dealing with diffuse pollution sources. At the community level, the negotiations about causes, responsibilities, and potential solutions for eutrified lakes can last for several months and/or years without any reconciliation between different sides (usually residents and homeowners with properties near the lake shore against the agricultural producers upstream) (e.g. Brooks et al. 2013; Schaefer 2014; Maclean 2017; Goswami 2018). And because funding for implementation of centralized wastewater treatment plants can be obtained through various government funding schemes (e.g. Canadian Council of the Ministers of the Environment 2006; European Court of Auditors 2015; The American Society of Civil Engineers 2017) when it comes to the decision regarding the type of wastewater treatment to be implemented (e.g. centralized wastewater treatment plants versus decentralized passive filtration systems for single or cluster houses) the communities by the lake shores will always vote in favour of the former.

The general absence of policy and regulations to address P discharges from residential, agricultural, and urban pollution sources, coupled with the current governance of innovative technologies for wastewater treatment, and the cost and time investments required for technology verification and implementation create insurmountable obstacles for the majority of inventors in placing their product/invention to the wastewater treatment market. As a result, there has been a very little aspiration to find/invent technologies and products for P removal from these non-regulated pollution streams (Drizo 2012). Consequently, despite the ever-increasing magnitude of HABs occurrences worldwide caused by P pollution, there is a significant gap in the wastewater treatment market when it comes to technologies and products for P removal from any of the non-regulated pollution sources.

## 3.2 P Removal from Municipal Wastewater Treatment Effluents (MWWTE)

### 3.2.1 Conventional Sewage Wastewater Treatment

Lofrano and Brown (2010) and de Feo et al. (2014) provided a comprehensive description of the history of sewage treatment and wastewater management. Although the evidence of indoor plumbing dates as far back as 8000 BCE (Ancientpages 2016), modern (conventional) sewage treatment practices (treatment of organic matter and domestic waste) began in the late nineteenth and early twentieth centuries.

The conventional wastewater treatment consists of a combination of physical, chemical, and biological processes with objectives to remove solids, organic matter, and pathogens. General terms used to describe different degrees of treatment are preliminary, primary, secondary, and tertiary and/or advanced wastewater treatment. The purpose of *preliminary treatment* is to remove coarse solids and other large materials often found in raw wastewater. This process typically includes coarse screening and grit removal. However, in most small wastewater treatment plants preliminary treatment does not include grit removal. The objective of *primary treatment* is the removal of organic and inorganic solids and floating materials. This occurs in a primary clarifier (sedimentation tank or basin) where liquids and solids are separated to facilitate biosolids settling at the bottom of the tank. Between 15% and 40% of the solids are removed at this stage, which takes about 4.5 hours. The solids are eventually pumped out of the tank and stored for later treatment (anaerobic decomposition in a digestion tank or incineration), and eventual use as a soil conditioner. The primary treatment acts as a precursor to secondary treatment. Approximately 25–50% of the incoming biochemical oxygen demand (BOD5), 50–70% of the total suspended solids (TSS), and 65% of the oil and grease are removed during primary treatment (Tchobanoglous et al. 2003; Topare et al. 2011).

The purpose of *secondary treatment* is to provide further treatment of the residual organics and suspended solids (SS) from primary treatment effluent, under either aerobic or anaerobic conditions. In general, aerobic treatment consists of intermittent sand filters or trickling filters, aeration, and settling tanks (e.g. activated sludge [AS] process) and oxidation ponds and aerated lagoons. The most common treatment under anaerobic conditions are anaerobic lagoons and digesters.

The purpose of *tertiary treatment*, also known as *advanced wastewater treatment* ('effluent polishing' stage) is to provide superior, final effluent treatment and remove constituents of concern including nutrients, toxic compounds, and pathogenic bacteria. It provides a final treatment stage with the aim to raise the effluent quality before it is discharged to the receiving environment (sea, river, lake, ground, etc.). This is the only treatment stage that provides nutrient removal.

### 3.2.2   Phosphorus Removal at MWWTPs

Development of technologies for phosphorus removal started in the 1950s along with the recognition and early evidence of eutrophication. Over the past half-century, a number of different methods have been developed each of them has different benefits and limitations. These involve a variety of physical (settling and filtration), chemical (precipitation, ion exchange, and adsorption), and biological processes (consumption by microorganisms or plants). Given that some of these methods for P removal and that MWWTE have existed for over a century, and most of them for over six decades, there is a myriad of wastewater engineering books, research papers, and reviews which describe these methods. For his reason only a brief description is provided here.

*Chemical Precipitation* is the most common method for P removal from municipal wastewater effluents (MWWE). The process was discovered in late nineteenth and early twentieth century (Wardle 1893; Wakeford 1913) and involves dosing of the effluent with calcium, iron or aluminium salts (e.g. Minton and Carlson 1972; Morse et al. 1998; Tchobanoglous et al. 2003; Ramasahayam et al. 2014; Ruzhitskaya and Gogina 2017; Bunce et al. 2018). However, despite its widespread use, this process has a number of limitations – the biggest one being that it creates large quantities of sludge which needs to be treated and/or disposed at exceedingly high costs. The quantity of chemicals required for the treatment depends on the number of factors including effluent P concentration, pH and alkalinity of the wastewater, the point of injection, and mixing modes (US EPA 2000b; Ruzhitskaya and Gogina 2017; Bunce et al. 2018). Recognizing the need for P reuse and recovery, a number of researchers started to investigate the potential re-use and recovery of P from sewage sludge (Chapter 4).

*Chemical Adsorption*: The use of industrial by-products and natural materials as P adsorbing media was been pioneered in the late 1990s along with the development of constructed wetlands (CW) technology and passive filters for P removal. A detailed history and description of media tested is provided later.

In recent years there has been increasing interest in investigations of magnetic *nanomaterial-based sorbents* for P removal due to their unique properties such as high surface area (generally between 1 and 100 nm) and easy solid–liquid separation by using external magnetic field as well as the potential for P recovery (e.g. Almeelbi and Bezbaruah 2012; Xu et al. 2016; Khodadadi et al. 2017). At present, the most extensively studied nanomaterials for P removal include iron (e.g. Almeelbi and Bezbaruah 2012; Yao-Jen et al. 2015; Khodadadi et al. 2017), activated carbon-silica (Al-Zboon 2017), aluminium-doped magnetic nanoparticles (Xu et al. 2016), titanium dioxide (Zheng et al. 2011). Lu et al. (2016) provided a comprehensive overview of different nanomaterials investigated for water and wastewater treatment. They underlined that despite promising results, a number of challenges need to be addressed before any wider commercial application. The most important being the potential toxicity of nanoparticles and their effects on human health and the environment. As the use of nanoparticles in water and wastewater treatment is an emerging field, current standards for assessing the toxicity of nanomaterials are insufficient (Li et al. 2015; Lu et al. 2016). Other disadvantages are related to the potential scaling up (e.g. the need for regular addition of nanoparticles) and inherited costs and energy requirements for their use (Lu et al. 2016). Therefore, despite increasing popularity and promotion via research funding to universities worldwide, to date, nanotechnologies have not offered better wastewater treatment and phosphorus removal compared to existing methods.

*Ion Exchange Technologies:* The ion exchange process was developed over a century ago and at the time was primarily used as a water softener. The main component of ion exchange is a microporous exchange resin, which is supersaturated with a loosely held solution. The most common applications of ion exchange resins are water softening (calcium and magnesium ions removal), water demineralization (removal of all ions), and de-alkalization (removal of bicarbonates). Cation exchange resins can also remove most positively charged ions (e.g. iron, lead, radium, barium, aluminium, and copper amongst others) whilst anionic exchange units can remove nitrate, sulphate, and other anions (Neumann and Fatula 2009). In recent years, the ion exchange method has gained popularity for phosphorus removal (e.g. Rittmann et al. 2011; Seo et al. 2013; Zarrabi et al. 2014). Some of the exchange materials used include metal-loaded chelating resins, hydrotalcites, iron-based hydroxide compounds, and capacitive deionization (CD) on electrodes (Rittmann et al. 2011). Although high P removal rates have been reported at laboratory scale, to date the implementation at the full-scale has been limited due to the costs and the sensitivity of some media to pH (Rittmann et al. 2011; Bunce et al. 2018).

*The crystallization method* is a solid–liquid separation technique, in which the solute crystallizes from the liquid solution and turns into a solid crystalline phase. The primary particle formation processes are based on seeding the wastewater in a fluidized bed to initiate nucleation or crystal birth which is followed by crystal growth until equilibrium (e.g. fluidized bed crystallization, FBC). This process has been developed and investigated for over three decades (e.g. Joko 1985; Eggers et al. 1991). Lu et al. (2017)

recently provided a comprehensive review of the crystallization methods and their applications including evaporation crystallization, cooling crystallization, reaction crystallization, drowning-out crystallization, and membrane distillation crystallization. Although the crystallization process is gaining in popularity (e.g. Binev 2015; Lu et al. 2017; Shokouhi 2017) due to its high P removal efficiency and potential for P recovery (Chapter 4), it has a number of challenges that limit its widespread use for P removal from wastewaters. Some of the major disadvantages include the mechanical complexity of operation, very high capital and operational costs, and a lack of reliable and long-term data (Lu et al. 2017).

*Electric coagulation* (EC) is electrolytic treatment based on the principles of electrochemistry and consists of electrodes that are arranged in pairs of two – anodes and cathodes (Föyn 1964). The cations produced electrolytically from anodes promote coagulation of contaminants from an aqueous medium thus providing treatment (Butler et al. 2011; Bouamra et al. 2012; Pulkka et al. 2014; Tian et al. 2017). The use of electricity for water treatment was first proposed in the UK in 1889 (Behbahani et al. 2011) electrocoagulation with aluminium and iron electrodes was patented in the US over a century ago (Dura 2013). However, due to high capital costs and the complexity of chemical coagulant dosing, this method did not gain interest until the 1970s (Mollah et al. 2004; Dura 2013; Nguyen et al. 2016). Over the past 45 years, the EC technology has been used for the treatment of a variety of wastewater effluents. Several comprehensive reviews of technology have been conducted describing different processes, advantages, and disadvantages (e.g. Mollah et al. 2004; Butler et al. 2011; Chaturvedi 2013). Mollah et al. (2004) highlighted EC technology benefits such as environmental compatibility, versatility, safety, selectivity, amenability to automation, rapid reaction, and the fact that the systems do not use chemicals and micro-organisms for water treatment. They also underlined the importance of EC cell reactor design in the technology performance. However, the technology also has a number of limitations the main being the electricity requirements and associated energy costs. Others are the occurrence of an impermeable oxide or passive film on the surface of the electrodes resulting in loss of the process efficiency and subsequent need for periodical replacement of the electrodes and high conductivity required which limits the process use for wastewaters containing low dissolved solids (e.g. Mollah et al. 2004; Dura 2013, Nguyen et al. 2016). Dura (2013) found that the uncertainty about the energy costs represents the major weakness to the more widespread use of EC over conventional methods.

### 3.2.2.1 Biological Treatment Methods

Biological removal processes rely on the phenomenon of 'luxury uptake' of phosphorus, which occurs when AS undergoes a period of anaerobiosis followed by aeration. The components of an AS plant are the aeration tanks (where biological oxidation takes place), settling tanks for the recovery of the AS, and piping and pumps system which serves to return the AS to the inlet end of the aeration tank (e.g. Horan 1990; Stensel 1991; Tchobanoglous et al. 2003). Biological removal processes in which P is removed in the sludge are referred to as *'mainstream processes'* and they are capable of achieving P effluent concentrations of 0.5 mg/l (Horan 1990). The phosphate-rich supernatant can be dosed with lime or other salts to precipitate phosphate (combination of biological and chemical processes) and then returned to the aerobic reactor where it is exposed to the additional luxury uptake of P. This option is known as the Phostrip process and represents an example of a 'sidestream process' (Horan 1990; Stensel 1991).

*Enhanced biological phosphorus removal* (EBPR) is the biological uptake of P by selected microorganisms known as the polyphosphate accumulating organisms (PAOs), which occurs in the AS process by recirculating sludge through anaerobic and aerobic conditions (e.g. Barnard 1975; Oehmen et al. 2007; Stensel 1991). PAOs are able to store phosphate as intracellular polyphosphate, and unlike most other microorganisms, they can take up carbon sources such as volatile fatty acids (VFAs) under anaerobic conditions, and store them intracellularly as carbon polymers, namely poly-b hydroxyalkanoates (PHAs) (Stensel 1991; Oehmen et al. 2007). Another phenotype of organisms that that store glycogen aerobically and consume it anaerobically as their primary source of energy for taking up carbon sources and storing them as PHAs are glycogen-accumulating organisms (GAOs). Oehmen et al. (2007) conducted a comprehensive review of both PAOs and GAOs including their identification and biochemical pathways. They also discussed the competition between the two groups of organisms as well as factors that affect them such as pH and temperature and dissolved oxygen concentration. In addition, they discussed in detail the Bardenpho processes ('BNR'), a complex system developed to achieve both nitrogen and P removal. More recently Bunce et al. (2018) reviewed biological P removal processes and discussed the latest advancements of the EBPR process. They also described up to date applications of EBPR including incorporation in membrane bioreactors (MBRs), granular sludge reactors, and sequencing batch biofilm reactors (SBRs). They highlighted that whilst the MBR process can achieve superior effluent quality, membrane fouling requiring a higher level of maintenance continues to be an issue, as well as the high capital costs.

Biological P removal can also be achieved in stabilization ponds. In these systems, oxygen is provided by natural surface aeration and algae during photosynthesis. P removal of about 80% had been reported during summer, however, during winter, the efficiency of these systems can reduce to 20% or less (Stensel 1991; Tchobanoglous et al. 2003).

### 3.2.3   Costs of P Removal in Municipal Wastewater Facilities (MWWTF)

Contrary to the vast amount of literature describing various methods and processes for P removal at MWWTFs, studies that investigated the cost-effectiveness of these processes is fairly limited (e.g. Jiang et al. 2004, 2005; WSDOE 2011; Falk et al. 2013; Ohio EPA 2013).

Jiang et al. (2004, 2005) conducted the first comprehensive analyses of the cost-effectiveness of several different P removal processes in MWWTFs in the USA, using two different approaches: (i) adaptation of existing facilities and (ii) construction of entirely new facilities. They considered a range of plant designs that would meet P discharge limits between 0.05 and 2.0 mg/l. When considering retrofitting and adaptation of existing MWWTFs they found that for an effluent TP concentration between 0.5 and 2.0 mg/l AS process followed by alum addition resulted in the most economical treatment for both 1 and 10 million gallons/day (MGD). For all investigated configurations, the costs of P removal to achieve discharge limits of 1 mg/l were in average fivefold higher for larger scale facilities (treating 10 MGD) compared to smaller ones (treating 1 MGD). Effluent treatment to reduce P discharge concentration by further 50% (from 1 to 0.5 mg P/l) resulted in 2.6–7-fold increase of treatment costs for both 1 and 10 MGD (Jiang et al. 2005). Overall they estimated that the costs of P removal to meet lower concentrations were about 150–425 $ USD/kg P removed. Falk et al. (2013) investigated the relationships and effects of different nutrient removal processes on greenhouse gas emissions at MWWTFs in the USA. Although they did not separate the costs of processes for P removal from nitrogen removal, their analysis suggested that a transition from a simple organic matter removal to nutrient removal would increase capital requirement by about 70%.

In 2011, Washington State Department of Ecology (WSDOE) contracted independent consultant, Tetra Tech, to conduct Technical and Economic Evaluation of Nitrogen and Phosphorus Removal at Municipal Wastewater Treatment Facilities for Washington State. They estimated that capital costs for WWTFs P removal upgrades to achieve TP discharge limits <1 mg/l would be $522 million/annually, whilst to achieve more stringent limits (below 0.1 mg/l) would be about $1555 million/year.

Most recently, Bashar et al. (2018) performed technical and economic evaluations of the high-performing P removal/recovery processes that have been implemented on full-scale systems such as: emerging secondary P removal processes; implementation of a side stream P recovery system in conjunction with a mainstream P removal process; and process modifications and their incremental operational costs to achieve a specific nutrient removal goal (effluent TP of 0.05 mg/l) using tertiary treatment. They compared overall annual economic cost (total capital, operational, and maintenance costs) and cost-effectiveness ($/lb of P removed) for six different full-scale treatment scenarios. These included (i) Modified University of Cape Town (MUCT) process consisting of a two-step anoxic process, (ii) five-stage (anaerobic, primary anoxic, primary aerobic, secondary anoxic, secondary aerobic) Bardenpho Process, (ii) MBRs, (iv) Integrated Fixed-Film Activated Sludge Systems with Enhanced Biological Phosphorus Removal (IFAS-EBPR), (v) MUCT process (mainstream P removal) coupled with side-stream P recovery from anaerobic digestate (struvite recovery), and (vi) tertiary media filtration (continuous backwash, upflow, deep-bed granular media filter and hydrous ferric oxide [HFO] coated sand filters). Analysis showed that electricity consumption accounted for the largest portion of the operating costs in each scenario, followed by sludge disposal in all scenarios (except P recovery in scenario 5). In addition, it was noted that for scenarios 1 and 6, sludge disposal could become prohibitively expensive if it has to be transported to distant locations. Chemical costs varied from 4% to 11% however, it was acknowledged that the costs of chemicals are highly variable. Overall, different treatment scenarios evaluated revealed the unit cost for P removal ranging from \$42.2 to \$60.9/lb. (\$93 to \$134/kg) of P removed. The MUCT combined with BNR tertiary reactive media filtration was one of the most cost-effective configurations (\$45/lb. P or \$99.2/kg P removed) producing effluent with a TP concentration of 0.05 mg/l. P reduction from the effluent due to P recovery as struvite was only 6% and as such could not aid in meeting stringent TP discharge limits. Chemical addition requirement was a major operational expense during the struvite precipitation process.

The EU funds allocated to wastewater infrastructure under the European Regional Development Fund (ERDF) and the Cohesion Fund were approximately 12.9 billion euros for the 2000–2006 programme period and increased to 14.6 billion euros for the 2007–2013 period (European Court of Auditors 2015). However, it is not clear what percentage of these funds is allocated specifically to MWWTFs upgrades for P removal or new treatment plants with tertiary treatment.

### 3.2.4 Novel Technologies for P Removal from MWWTFs

Most of the MWWTFs located in the countries that have established requirements for P removal from municipal wastewater effluents have set total P discharge limits of 1 mg/l and 2 mg/l. However, given the continuous increase of areas affected by eutrophication across Europe, the EU Water Framework Directive (WFD) started to consider more stringent P limits of 0.1 mg/l (for larger treatment works) and 0.5 mg/l (for smaller treatment works) (Valley 2016). Consequently, in the past few years, finding novel technologies capable of meeting the new P consents became a priority for a number of leading water and sewerage companies in the UK (e.g. AquaEnviro 2016, 2018; Valley 2016; WWTonline 2016a; UKWIR News 2017). The UK Water Industry invested 50 million pounds to support novel trials for solutions for P removal by major water and sewerage companies (UKWIR 2017). A UK corporation Severn Trent has spent £120 million on additional P removal trials (Valley 2016).

In October 2016, several new products, technologies, and processes capable of meeting stringent P removal limits were presented by the leading water and wastewater treatment corporations at the 10th European Wastewater Management (EWWM) Conference and Exhibition held in Manchester, UK (AquaEnviro 2016). These included ACTIFLO® high rate clarifier exclusively developed and patented by Veolia Water Technologies (Veolia 2018), FilterClear™ a high rate multimedia filtration system developed by Bluewater Bio (Bluewater Bio 2018), cloth media filters tested by Severn Water Trent (WWTonline 2016b), Greenleaf filters by Suez Advanced Solutions UK, Tertiary Air Flotation by Njihuis H2OUK (AquaEnviro 2016). Phosphorus removal processes to achieve stringent P discharge consents at MWWTFs remained one of the key topics of the EWWM Conference in 2018 (AquaEnviro 2018). Few new processes were introduced, such as Soneco® which combines electrolysis with ultrasound developed by KP2M Ltd., trading as Power & Water (P&W). The electricity is used to stimulate coagulation and flocculation and creation of metal hydroxide networks that adsorb colloids, fines, clays, metals, and nutrients. The resultant floc particles are then separated downstream in solid–liquid separation treatment step. Ultrasound is applied during electrolysis with the aim to increase the number of nucleation sites for flocculation (WWTonline 2018).

The novel P treatment processes described above are still in the early stages of investigation and the cost-efficiency of their potential implementation on MWWTFs in the UK and across the EU has not been reported. Given their complexity, it is likely that treatment will be much more expensive compared to traditional methods. Moreover, with P pollution originating from all other sources being ignored, the question remains on whether P regulations focus should remain on developing and requiring stringent P discharge consents for MWWTFs. P removal at the source could provide not only the greatest reductions/recovery of P (Chapter 4) but also decrease risks and costs of processes and infrastructure downstream (Bullen 2017).

## 3.3 Phosphorus Removal from Residential Wastewater Effluents (Onsite Residential Wastewater and Disposal Treatment Systems)

The main approaches for P reduction from residential (individual or cluster) households' wastewater effluents include source reduction, source diversion, and septic tanks with subsequent soil treatment. The most common source reduction methods are the reduction or elimination of P use in domestic products (e.g. household waste reduction and using P-free detergents). Source diversion methods include systems for wastewater separation at the source to urine ('grey water') and/or faeces ('black water', which contains the bulk of the nutrients present in wastewater). This concept, known as Ecological sanitation (EcoSan), was developed in Sweden and Germany over 20 years ago and has mostly been applied in developing countries (Winblad and Simpson-Hebert 2004).

However, the most typical and frequently applied treatment technique worldwide are *onsite septic tanks* followed by a soil absorption field. This type of treatment is also known as a subsurface wastewater infiltration

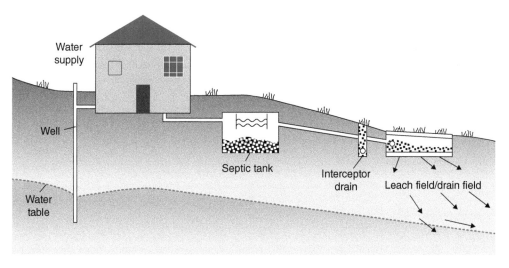

**Figure 3.1** Typical layout of an onsite septic system. Source: Environmentalenhancements (2018).

system or soil treatment and disposal field ('leach field/drain field'), and is commonly referred as OSS (Figure 3.1) (Beal et al. 2005; Gill 2011; Withers et al. 2012).

Septic tanks require regular inspection and maintenance in order to ensure the system is well-managed and functioning properly. However, due to homeowners' negligence or lack of knowledge, OSS are often not properly managed, resulting in malfunctioning. a conservative estimate for the USA is that OSS fail at the rate of 10–30%. In addition, over 50% of onsite systems are over 30 years old, seriously increasing their likelihood of failing (Indiana University Library 2015). Studies conducted in the UK a decade ago estimated that 80% of the 1.5 million OSS in the UK are working inefficiently with respect to P discharges (Brownlie 2014). In New South Wales, Australia, as much as 70% of OSS are reported to fail in meeting public health protection and environmental standards (NSW Department of Local Government 2000).

### 3.3.1 Potential Phosphorus Loading from OSS

The median concentration of P in septic tank effluent is about 8–10 mg/l (Dudley and May 2007; Toor et al. 2011; Pipeline 2013). A single household (three bedrooms) in the USA discharges 1700 l/d (450 gal/d), whilst in Europe approximately 900 l/d resulting in 6.2 kg P (USA) and 3.3 kg P (Europe) discharges of P annually. Taking into account the fact that on average 26–50% of the population is not connected to wastewater treatment plants, the OSS discharge significant quantities of P to both fresh and coastal waters worldwide (Table 3.1). Moreover, freshwater lakes are often surrounded by seasonal cottages and commercial developments. The number of incidences of harmful algae blooms have been increasing at alarming

Table 3.1 The potential daily and annual phosphorus loading from onsite septic tank effluents.

| Country/ Region | Population | Number of septic systems[a] | Daily Phosphorus loading (ton/d) | Annual Phosphorus loading (ton/year) |
|---|---|---|---|---|
| USA | 325 700 000 | 28 227 333 | 479.9 | 175 150.6 |
| Canada | 36 095 000 | 3 128 233 | 53.2 | 19 410.7 |
| European Union | 512 600 000 | 44 425 333 | 399.8 | 145 937.2 |
| Europe[b] | 14 273 332 | 1 237 022 | 11.1 | 4063.6 |
| Europe[c] | 19 722 409 | 3 287 068 | 29.6 | 10 798.0 |
| Europe[d] | 300 149 569 | 50 024 928 | 450.2 | 164 331.9 |
| Europe Total | 846 745 310 | 98 974 351 | 890.7 | 325 130.7 |
| Australia | 24 770 000 | 2 146 733 | 19.3 | 7052.0 |
| New Zealand | 4 749 598 | 411 631 | 3.7 | 1352.2 |
| TOTAL | | | 2337.5 | 853 227 |

[a] Based on the conservative estimate that only 26% of the population is served by onsite septic systems (OSS) in most countries.
[b] The European Economic Area (EEA) but not members of EU: Iceland, Norway, Liechtenstein, and Switzerland.
[c] Western Balkans: Albania, Bosnia and Herzegovina, Serbia, Montenegro, Kosovo and Macedonia, in average 50% of population is served by OSS.
[d] Russian Federation, Ukraine, Azerbaijan, Belarus, Georgia, Moldova, Armenia, Turkey, also have about 50% of population served by OSS.

rates, causing significant losses to tourist and real estate industries. Today, over 50% of freshwater lakes in North America and Europe are being affected by eutrophication. An increasing number of studies showed that P pollution from OSS supplements, and may exceed the pollution from agricultural operations, in particular in watersheds where on-site farm conservation practices are widely implemented (e.g. Withers and Jarvie 2008; Withers et al. 2009, 2012, 2014a; Jarvie et al. 2013). However, despite increasing scientific evidence on P discharges from OSS, to date, Scandinavia remains the only region in the world that has developed policy driven regulatory requirements for P reduction from this source (Chapter 2). Consequently, P discharges from OSS remain untreated worldwide (Box 3.1).

Box 3.1  Annual Phosphorus Loading from Onsite Septic Tank Effluents

The estimated potential annual P loading from septic tank effluents (based on the percentage of population served by septic tanks and occupancy of three people per house) for USA, Canada, Europe, Australia, and New Zealand is nearly 1 million ton P/year. Taking into account the fact that approximately 50–60% of the population in China, Japan, India, and South America is not connected to centralized wastewater treatment plants, contributing with 4500 ton of P daily, the total annual input of P from septic systems effluents and residential wastewater effluents is over 2 500 000 tons P/year.

Given that P removal regulations for OSS currently exist only in Scandinavia, just 0.2% (5904.7 tons P/year) of the world P annual loading or 1.8% of the EU and EEA combined P loading from this source is currently being addressed.

### 3.3.2   Mitigation of P Pollution from OSS

Mitigation of P pollution from OSS is impeded by several major factors including (i) the traditional misconception that soils in a drain field and beneath will effectively and continually retain phosphorus over long periods of time; (ii) a general lack of data on OSS locations and functionality; (iii) difficulties in the field measurements of P discharges from OSS; (iv) literature discrepancy regarding the extent of P contamination from OSS effluents; (v) regulators and home-owners concerns over the costs of OSS retrofitting with P removal products (in a case where P discharge standards and regulatory requirements for P reduction were to be developed). Some of these factors are discussed in the following paragraphs:

1   *The traditional misconception regarding drain field soils effectiveness in P retention:*
   On average, about 20–30% of P is expected to be removed in a septic tank through solids settling in well-managed systems. The remaining P is assumed to be effectively captured and retained by the soils in a drain field and beneath through chemical processes of precipitation/ adsorption/complexation with iron ($Fe^{+2}$ and $Fe^{+3}$), aluminium ($Al^{+3}$), and calcium ($Ca^{+2}$) cations (Brady and Weil 1996; US EPA 2002; Pipeline 2013). Thus, OSS are considered an effective and permanent solution for treatment of domestic households wastewater in rural areas (Gold and Sims 2006; Pipeline 2013; Withers et al. 2014a). However, phosphorus retention by soils is a finite process (Richardson and Craft 1993; Brady and Weil 1996, Drizo et al. 2002) and cumulative addition of large quantities of P to drain field soils will significantly decrease their effectiveness in P retention over time. This has been corroborated by a number of scientific studies over the past 10 years which revealed that OSS can make significant contribution to P loading of surface and ground waters and subsequent eutrophication (Dudley and May 2007; Withers and Jarvie 2008; Neal et al. 2010; Withers et al. 2012; Eveborn 2013; Brownlie 2014; May et al. 2014; Withers et al. 2014a).
   The extent and capacity of soils to retain P is affected by many factors including their mineralogy, type, physico-chemical properties, climate, and land use management. Volcanic soils (Andisols) have the greatest P-sorption of all soils due to a high content of amorphous material, yet they cover only 1% of earth surface (Neall 2013). Highly weathered soils (such as Oxisols and Ultisols) also have high P retention capacities due to the presence of large amounts of aluminium and iron oxides and highly weathered kaolin clays (Brady and Weil 1996; Mnthambala et al. 2015). However, oxisols occupy about 7% of the nonpolar continental land area on Earth, located mostly in the equatorial regions of South America and Africa (*Encyclopedia Britannica* 2018a). Ultisols cover just over 8% of the nonpolar continental land area on Earth and are found in humid temperate or tropical regions, including the southeastern USA

and China, and in the humid tropics in South America and Africa (*Encyclopedia Britannica* 2018b). Thus, soils with the highest P retention capacity (PRC) cover less than 20% of the land surface. Physico-chemical properties of soils such are porosity, texture, hydraulic conductivity, pH, the oxidation/reduction status and their cation exchange capacity, organic matter content all affect soil adsorption capacity (Richardson 1985; Richardson and Craft 1993; Drizo et al. 1999). Sites with impermeable soils, high clay content, or shallow bedrock or those with slopes greater than 15% will have little capacity to attenuate P (Ready 2008; Pipeline 2013). Sandy soils are most susceptible to P leaching.

In addition to site-specific conditions, land use and management and other human factors such as hydraulic overloadings, garbage overloading, and excessive use of chemicals, often lead to OSS failures (Ready 2008). OSS regulations stipulate rules about minimum distances between a septic system and any surface water, groundwater, and foundation drains to prevent flooding the leach field. Nonetheless, malfunctioning OSS often lead to excess P loading to ground and surface waters, in particular, when located in headwater catchments or in the close proximity of lakes and coastal shorelines (e.g. US EPA 2002; Miller and Wright 2014; Withers et al. 2014a).

2 *OSS locations, numbers and functionality data shortage:*
In the USA, the US Census Bureau field staff collect data on up-to-date housing statistics every two years. Thus the Bureau provides some general statistical information such as the percentage of OSS located within Standard Metropolitan Statistical Areas (cities and their suburbs). It also provides information about the percentage of the total and individual states' population served by the OSS. The highest percentage of population served by the OSS (50–55%) is found in the states of Vermont, Maine, New Hampshire, and North Carolina, followed by Alabama, South Carolina and Wyoming (40%). An average value for the USA population served by the OSS is 25%; in addition to households, about 33% of new construction is served by onsite systems (Indiana University Library 2015). Whilst the knowledge on the percentage of the population using OSS is available, the actual information on the OSS locations within the watersheds, individual states, or at national level has been lacking with the exception of Georgia, Illinois, and Ohio. The University of Georgia Marine Extension Centre developed a comprehensive geographic information system (GIS) inventory for the coastal zones (Walker et al. 2003). States of Illinois and Ohio also used GIS for septic systems mapping in certain counties. However, these are isolated cases as the lack of funds prevented wider application and development of OSS inventory across the USA (Rountree 2014).

Recognizing the potential for serious environmental degradation from older OSS, the US EPA (2002) developed the Onsite Wastewater Treatment Manual, which describes several models and approaches for their risk assessment including a subjective vulnerability assessment, a probability analysis of water resources impact from wastewater discharges, and contaminant transport modelling. In addition, they described a DRASTIC model which was developed by the US Geological Survey to rate groundwater vulnerability using weighted factors of hydrogeological settings (US EPA 2002; Kinsley and Joy 2006).

In the EU, the Groundwater Directive 2006/118/EC was revised in 2006 (European Parliament and Council 2006) and set criteria to control pollution discharges from OSS. To reflect this change, a need for registration of OSS in England and Wales and environmental permits for those located in areas vulnerable to groundwater pollution was considered in 2010 regulations (May et al. 2010; Brownlie 2014). Today, OSS registration is legally required in Wales. In Scotland, owners are obliged to register their OSS with the Scottish Environment Protection Agency (SEPA), under the Water Environment Controlled Activities (Scotland) Regulations (2011), but this is only legally imposed if the property is to be sold (Brownlie 2014). However, the requirement for septic tank registration in England was removed on 1 January 2015 (Bennett et al. 2014; Wentworth 2014). This imposed even greater problems in identifying the numbers, locations, and functionality of OSS. The official data state that there are only 200 000 OSS in the UK, whilst the works of May and co-workers have demonstrated that there are 1.4 million OSS (May et al. 2014). Using aerial photography and large area statistics, May et al. (2014) provided evidence that in some catchments in England nearly 95% of OSS are not registered and are thus excluded from the official reports which often state that P pollution from OSS does not represent a threat to water resources. Therefore, obligatory OSS registration is essential to gain a better understanding and to elucidate the facts about their contribution to P pollution and eutrophication of lakes, river catchments, and coastal areas. Obligatory OSS registration would provide base data necessary for the development of interactive maps showing their exact locations, and potential excess P loading (similar to those developed in Georgia, Illinois, and Ohio). If the European Environment Agency (EEA) were to develop an information database on OSS locations amongst Member States, and generate maps similar to those made for urban wastewater treatment compliance (EEA 2015) it could greatly enhance the current state of knowledge regarding P pollution sources in the European waters, as well as the innovation in P removal and recovery from these sources (Chapter 4).

3 *Literature Discrepancy:*
OSS contribution to eutrophication has been a subject of numerous studies over the past several decades. Nevertheless, there is still a significant discrepancy in assessing their role as a potential threat to surface and groundwater quality. For example, the US EPA (2002) recognized OSS, and in particular the older ones, as the primary cause of eutrophication in thousands of inland lakes in the USA. However, Gold and

Sims (2006) reviewed data from over 20 selected studies published during a 25 year period (1976–2000) in Ontario province of Canada and several states in the USA (Wisconsin, Virginia, Massachusetts, Washington, and New York) and concluded that the risks of P contamination of wells or surface waters from septic tank outflows are very limited. Amongst reviewed studies, they cited Gilliom and Patmont (1983) who studied eight septic systems located at Pine Lake, Puget Sound, Washington and concluded that it is very unlikely that more than 1% of P from the septic tank effluents would reach the lake. Whilst this may have been a case 35 years ago when the original study was conducted, today, the eutrophication of the Spokane county of shorelines and lake waters (including Pine Lake) has reached such an alarming proportions that they considered development of a P discharge standard (HDR 2007; Miller and Wright 2014; E. Miller, personal communication). Similarly, solutions to P loadings from septic tank effluents are being sought at Chautauqua Lake, Long Island and other watersheds throughout NY State (Diers 2013).

Even the leading North American authority on onsite wastewater treatment systems, the State Onsite Regulators Alliance (SORA) has made misconceptions in their Report on the Onsite Wastewater Phosphorus Regulation (SORA 2012). SORA's mission is to advance the field of knowledge and practice of those who regulate onsite wastewater programmes by increasing awareness of the latest technology, research, environmental health issues, and new federal initiatives that will affect the decentralized wastewater industry (SORA 2012). Its members include government regulators of decentralized and onsite wastewater systems from all 50 states, US Territories, Native American Tribes, and Canadian Provinces. However, their 2012 Report states that 9 out of 50 States have existing regulations that require P removal from OSS (Figure 3.2), whilst P regulations do not exist it any of these States (M. Degan, personal communication; Vermont Agency of Natural Resources 2012; Drizo 2013; Miller and Wright 2014; Rountree 2014). The fact that errors were made by the SORA impedes the future assessments of the OSS contributions to P loading even more.

### 3.3.3   Phosphorus Removal Methods and Technologies for OSS Effluents Treatment

The focus of this section is on the methods and technologies for onsite sewage treatment (single or cluster houses) that can be easily integrated with the existing infrastructure (Figure 3.1). These include

I   Chemical precipitation with Ca, Fe, or Al salts and/or organic polymers and

II   Commercial products and technologies (passive filtration technologies and products based on electrochemical processes).

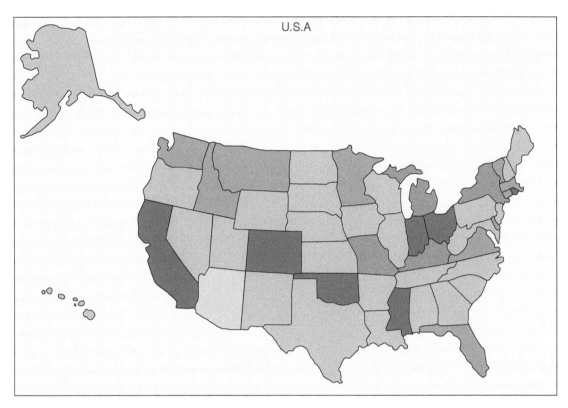

U.S.A

**Figure 3.2** Status of Regulation of P from Onsite Wastewater Systems in the USA. States shaded light blue are those for which P discharges from OSS were not considered to be a problem at this time; States shaded dark blue are those for which respondents indicated that P discharges were problems in some areas but regulations were not being considered. Purple shaded states are those for which respondents indicated that P regulations are being considered. Green shaded states are those for which regulations for P from OSS exists. Withers et al. (2014a).

### 3.3.3.1 Chemical Precipitation

The chemical precipitation of P with iron or aluminium salts and/or organic polymers method has been adopted from traditional engineering in centralized municipal wastewater treatment and MBRs (Section 3.2). The chemical can be dosed either after the biological treatment step, at which point the water is conveyed to a sedimentation tank or added at an earlier stage (e.g. in the house) with subsequent precipitation of P in the septic tank. However, similarly to MWWTP applications, the addition of chemicals results in increased sludge production. Thus, the septic tank would need to be pumped out two to three times a year on average, considerably increasing the cost of treatment. In addition, precipitants used may contain contaminants such as heavy metals which can prevent the land application of produced sludge (Vidal et al. 2018). Another disadvantage of this technique is that it requires continued maintenance. In addition, the regular restock of the chemical can be costly, time-consumingm and tedious for the owner. Membrane-based technologies such as MBR have been suggested being suitable for on-site wastewater treatment (Fane and Fane 2005). Data suggest that they can remove P to 65% (Boehler et al. 2007) and 70–90% under certain conditions (Abegglen et al. 2008). However, they require intensive monitoring via remote control over the internet and regular servicing to prevent malfunctioning, which requires considerable and continuous financial investments from the owner (Abegglen et al. 2008).

### 3.3.3.2 Commercial Products and Technologies for OSS Effluent P Treatment
#### 3.3.3.2.1 *Passive Phosphorus Filtration Systems*
##### 3.3.3.2.1.1 History and Background – Media Selection

The use of industrial and natural by-product materials to enhance P removal was pioneered in the 1990s by several scientists from Europe (Drizo et al. 1997, 1999; Johansson 1997; Zhu et al. 1997) and Australia (Mann 1997). Building on the concepts from traditional wastewater engineering (Section 3.2), they started investigations of locally available and inexpensive materials rich in Ca, Fe, and Al oxides content. Furthermore, Drizo et al. (1999) set out several additional criteria for the materials selection which included a range of chemical and physical properties such as PRC, hydraulic conductivity, cation exchange capacity, porosity, and particle sizes. These investigations established the foundation for the new research field on the industrial and natural by-products use for phosphorus removal from wastewaters. During the past two decades over 250 industrial by-products and natural materials were investigated in laboratories around the world. Their performance has been described in over 2000 research papers (e.g. Baker et al. 1998; Sakadevan and Bavor 1998; Johansson 1999a,b; Drizo et al. 1997, 1999, 2002, 2006; Brooks et al. 2000; Brix et al. 2001; Forget 2001; Molle et al. 2003; Brogowski and Renman 2004; Adam et al. 2005; Johansson-Westholm 2006; Renman and Renman 2010). Drizo and co-workers tested over 80 different natural and industrial by-products (Drizo et al. 1999, 2002, 2008b; Forget 2001; Drizo 2012). More recently, Penn et al. (2011) started investigations of industrial by-products for P removal and introduced the term 'phosphorus sorbing materials (PSMs)'. Loganathan et al. (2014) reviewed 30 different materials focusing only on materials that can remove P via sorption process and including emergent mineral and synthetic materials, such are layered double hydroxides (LDH), zirconium hydroxide, iron–zirconium binary oxide, iron–manganese binary oxide, metal oxide–polymer mixtures. Comprehensive reviews of this research were conducted by several researchers (Johansson-Westholm 2006; Chazarenc et al. 2008; Cucarella and Renman 2009; Vohla et al. 2011; Klimeski et al. 2012; Loganathan et al. 2014). Whilst these studies were very important in advancing the scientific knowledge on the types of materials that could have potential in removing P from domestic wastewater they have several shortcomings, the most important being:

1 *Use of an inadequate method for determination of maximum PRC of the material*:
  In all of the above studies, without an exception, P adsorption capacity of candidate materials was measured in the laboratory employing

'pseudo-equilibrium' batch experiments – a technique developed and established as a method of measuring P retention characteristics of soils (Langmuir 1918; Olsen and Watanabe 1957) and not for aggregate materials (Drizo et al. 2000; Forget 2001; Drizo et al. 2002). The procedure, known as the '24 h Standard batch measurement method' consists of making five different solutions containing a range of P concentrations (e.g. 'initial P concentrations'), submerging the potential candidate material into the solutions and subjecting it to vigorous shaking for 24 h or more. The material's PRC is then calculated as a difference in P concentration in the solution before and after shaking and fitting the experimental data to Langmuir or Freundlich isotherm equations. The former one is used much more frequently because it allows 'determination' of so-called maximum PRC. The PRC of the material is considered the most important engineering parameter as it is used for the estimations of a full-scale system longevity (if implemented in the field). However, the 24 h Batch method employs eight different variables, including the range of initial P concentrations, material weight, equilibration vessel used, material weight to solution ratio, head-space, type of microbial inhibitor – if used, type of shaker and the rotational speed of a shaker (Nair et al. 1984). All of these variables affect the materials' capacity to retain P and consequently the 'maximum' value derived from the Langmuir equation. Moreover, Forget (2001) and Drizo et al. (2002) demonstrated that PRCs of materials derived from P batch experiments using Langmuir equation do not provide realistic values and should only be used with caution for estimating the longevity of full-scale systems. They also suggested that laboratory investigations of potential materials ought to be coupled with longer-term column experiments. They further pointed out that the latter one would enable the determination of a P saturation potential, a parameter which provides more realistic and reliable estimates of the full-scale systems life-span (Drizo et al. 2002). However, to date, there had been no consistency in laboratory methods used for the determination of materials' PRCs, which greatly hinders comparisons of their performances and potential for use in field applications. And yet, despite the lack of a reliable method researchers continue to use the Langmuir equation for laboratory investigations of materials' P retention capacities (e.g. Penn et al. 2011; Vohla et al. 2011; Klimeski et al. 2012; Sellner et al. 2017).

2   *Luck of investigations with real wastewater effluents:*
    Along with the increased recognition of the need to find inexpensive methods to remove P from other pollution sources (e.g. agricultural wastewater and runoff and urban stormwater runoff), the number of studies that employed column experiments for determination of materials' PRCs has also grown. Nonetheless, the studies using real wastewater effluent remained limited (Weber et al. 2007; Drizo et al. 2008a); Lee et al. 2010; Bird and Drizo 2010; Jenssen et al. 2010; Renman and Renman 2010; Vohla et al. 2011; Klimeski et al. 2012, Vidal et al. 2018).

One of the reasons may be the absence of regulatory requirements. Other reasons may be associated with challenges in setting the experiments with real wastewater effluents such as the collection, transport, and disposal of large volumes of wastewater.

To date, only a few materials have been developed into products/technologies for commercial use in decentralized wastewater treatment. These include opoka (branded as Polonite systems offered by Ecofiltration Nordic AB (formerly known as Bioptech AB), Sweden; Steel slag aggregates (SSA) (PhosphoReduc systems, offered by Water and Soil Solutions International in USA and Europe); combination of SSA with limestone or limestone and gravel (Phosphex and PhosRid systems, Canada and USA) and light expanded clay aggregates (Filtralite P, Norway).

### 3.3.3.2.2 European Onsite Wastewater Treatment Market

As the first region in the world that developed and enforced P discharge and removal regulations for OSS over 10 years ago (Chapter 2), Scandinavia is currently a leader in the number of products and methods (commonly referred to as 'P traps') for onsite wastewater effluent P removal and treatment.

The major commercial products that are available and promoted as 'P reactive media' or 'P reactive filters' on the Scandinavian market are Polonite® (Ecofiltration 2018) and Filtralite® P (Saint-Gobain Weber, Norway).

### 3.3.3.2.3 Polonite Systems

Sweden has approximately 1 million single and individual OSS. Of these, approximately 400 000 are not up to regulation standard, whilst a very small portion of them has some kind of phosphorus trap or treatment (Eveborn 2013). Every year approximately 11 000 new individual OSS wastewater units are installed in new or old houses. They are all required to have P removal devices. Polonite systems and reactive media have been most popular, with estimated 5000 treatment units installed in the past 10 years (Ecofiltration 2018). They hold approximately 60% of the Swedish market. Research has also shown that there is a potential of spent Polonite media re-use as a fertilizer (Cucarella 2009).

Polonite reactive media is made by extraction of a natural resource, a Ca rich bedrock opoka, found only in Poland, Russia, and Ukraine. Opoka is a Ca rich sedimentary deposit with moderate to high P-sorption capacity. It consists mainly of $SiO_2$ and $CaCO_3$ but also contains $Al_2O_3$ and $Fe_2O_3$ (Brogowski and Renman 2004; Renman and Thilander 2006; Cucarella et al. 2007). Brogowski and Renman (2004) found that P-sorption efficiency of opoka increased significantly when heated over 900 °C, due to the transformation of $CaCO_3$ into CaO. The new material derived from a particular type of opoka heated at high temperatures has been branded as Polonite. Until 2006 material was manufactured by the major Swedish construction company NCC AB, and from 2006 to 2016 by Bioptech AB. In 2016, Bioptech

AB changed its name to Ecofiltration Nordic AB. Over the past 12 years considerable research on Polonite media for P removal from wastewaters has been carried out at the Royal Institute of Technology KTH (Cucarella and Renman 2009; Renman 1998; Eveborn 2013; Nilsson et al. 2013) led by Professor Gunno Renman, the co-inventor of the compact filter bed technology (Renman and Thilander 2006).

Polonite-based P removal systems are offered by Ecofiltration Nordic AB. The two main products are:

i    Single household treatment unit, housed in a plastic tank to serve as an add-on system to a septic tank. The pre-treated wastewater is fed to a biological treatment with supplemental aeration for the removal of organic matter (90%) and total nitrogen (TN) (50%). A bag filled with Polonite (500 kg, 720 l) and designed in upflow hydraulic regime is placed in the middle of the tank. Lifespan of material is two years achieving P removal efficiency of 90%, decreasing P concentration at the discharge to <1 mg/l. A lid opening facilitates easy filter bag replacement. The unit has a low and competitive price of SEK 65 000 (6314 €). However, the system requires media replacement every two years if the wastewater load is 800 l/d, at a cost of SEK 7900 (767 €). The lifetime of a system is estimated to be 30 years and if the fee for management is included the annual cost will be approximately 700€.

ii   Polonite bags (500 kg), having a lifespan of ~2 years sold at SEK 9000 (874 €) each.

     The main advantages of Polonite systems are that they have a proven performance of high P removal efficiency. The material is light in weight, thus having a relatively low cost of transportation. It also has a potential for re-use as a P rich soil amendment and fertilizer and thus provides an opportunity for P recycling and re-use. However, Polonite systems have several significant disadvantages: (i) they are based on natural material obtained by the extraction of bedrock located in regions outside Sweden (mainly Poland, though there are opoka reserves in Russia and Ukraine); (ii) material manufacturing requires thermal treatment, which is naturally high energy consuming and may contribute to green-house gas emissions (GHG), with additional emissions created during transportation from the point of extraction to the manufacturing facilities and sites location; (iii) like all alkaline materials, Polonite generates highly elevated pH effluents, often exceeding 12 (Eveborn 2013; Nilssen et al. 2013). In addition, they require frequent replacement of the media (Table 3.2).

iii  Polonite Ecobox, offered since 2010 by Skandinavisk Ecotech AB exclusively in Sweden and Finland (K.-G. Niska, personal communication). Ecobox is specially designed to serve PE ranging from 5 to 80 pe (K.-G. Niska, personal communication; https://www.ecot.se). About 3500 systems have been installed to date (Figure 3.3).

**Figure 3.3** Example of Polonite P traps installation in Sweden. Source: Mr. Karl-Gustav Niska, Skandinavisk Ecotech AB.

Similarly to polonite bags, the system can achieve P reduction of over 90%, media replacement is every two years, and the system generates pH effluent of over 12. Initial capital cost are 72.900 SEK (6950 €).

### 3.3.3.2.4 Filtralite P

Filtralite P are light-weight aggregates (LWA) originally developed to enhance PRC of the Norwegian Light Expanded Clay Aggregates (LECA) (Adam et al. 2007; Cucarella and Renman 2009). They are produced by heating a mixture of illite rich marine clay soil and dolomite to 1200 °C, expanding it to form porous particles. Its grain size is in the range 0–4 mm, resulting in high specific surface area, effective porosity of 40% and dry bulk density of 550 kgm$^{-3}$ (Adam et al. 2005). The material contains only 3.1% Ca, 2% Al, 0.6% Fe, and 0.7% Mg (Adam et al. 2007) and it had shown good potential for P removal in several laboratory batch and column studies, indicating maximum P sorption capacity of 3.3 g P/kg (Adam et al. 2005, 2007). However, field studies using Filtralite P as a substrate for

**Figure 3.4** Filtralite®
P system installation in
Norway. Source: Professor
Peter Jenssen, Norwegian
Agriculture University.

CW revealed significantly lower P sorption capacity, ranging from 0.15 g
P/kg to 0.4 g P/kg after a few years of system operation (Adam et al. 2007).
Jenssen et al. (2010) reviewed the performance of Filtralite P for both
P removal and its potential re-use as a fertilizer, implemented in nine full-
scale systems in Norway (Figure 3.4), Sweden, Finland, and Denmark.
Each system consisted of a septic tank, a pump well, a vertical flow single
pass aerobic biofilter (filled with lightweight aggregate 2–10 mm particle
size), a subsurface horizontal flow filter filled with Filtralite P and an outlet
well, resulting in a total treatment volume of 40 m³/single household
(Jenssen et al. 2010). The results showed that Filtralite P maintained discharge P concentrations below 1 mg/l after three years of system operation.
The material also showed potential for vegetation (ryegrass) growth indicating that it could be used as a P fertilizer. However, using Filtralite P for
single household P treatment has several disadvantages: (i) they require
investment costs of €15 000 with additional €5000 or more for material
replacement after it reaches saturation (estimated at 15 years, based on
laboratory studies); (ii) Filtralite P has high costs (about 30% of a total
system cost); (iii) similar to Polonite systems it generates effluents with
highly elevated pH (12–13); (iv) it is manufactured by heating clay at high
temperatures and thus contributes to GHG emissions.

Filtralite P was developed in collaboration between the LECA Company
and the Norwegian Agriculture University about 25 years ago (T. Krogstad,
personal communication). Originally it was manufactured and supplied by
Maxit LWA and Optiroc (Optiroc 2003; Adam et al. 2005). Since 2010
Filtralite P is manufactured by St Gobain Weber, the leading premix and
expanded clay producer and provider of premix plants and machinery in

Europe with more than 100 plants in 30 countries (Weber St Gobain 2018). However, despite its presence on the wastewater treatment market for over 10 years, to date, only a few hundred Filtralite P systems have been installed for the treatment of domestic wastewater. Commercial systems for P removal from domestic wastewater were installed mainly in Norway and Denmark and some in Sweden and Finland. Filtralite P media was also pilot tested on several locations in the USA, however, it has not been employed in commercial applications. Some of the pilot projects showed that Filtralite P loses its P sorption capacity after a short period of time and is prone to clogging (Biomicrobics, personal communication).

## 3.4 North American Onsite Wastewater Treatment Market

### 3.4.1 Phosphex™

The Phosphex treatment system was invented by research scientists Blowes, Ptacek, and Baker at the University of Waterloo, Canada in the late 1990s. US, Canadian, and UK patents (Blowes et al. 1996) have been issued for the technology (University of Waterloo 2018). Originally the system was developed for the treatment of various contaminants (Blowes et al. 1996) via media that consists of the combination of metal oxide media and hydroxides of iron and other materials, and was named Wmetox. The invention further states that Wmetox ought to contain at least 5% of calcium oxides and 15% of iron oxides and can contain limestone and basic oxygen furnace slags, the later composing at least 25% of the total Wmetox weight.

The University of Waterloo Phosphex system can be installed as a horizontal reactive barrier below a septic system tile field, a vertical barrier located in the pathway of horizontally flowing contaminated water sources, or within an enclosed treatment container (University of Waterloo 2018). Whilst it has shown efficient P removal in laboratory and field trials (Baker et al. 1998), its application in the field revealed significantly smaller P retention (B. Eichinger, personal communication). In addition, similarly to other alkaline based materials (e.g. Polonite and Filtralite P), the Phosphex system generates effluents with highly elevated pH (>12). Problems with systems clogging have also been experienced (B. Eichinger, personal communication). Systems are housed in concrete tanks and are offered at the cost of ~12–15 000 US$/single household (B. Eichinger, personal communication). There is no indication on the frequency of the media replacement which requires additional costs (University of Waterloo 2018).

Since its inception, there have been several attempts to licence the technology to Canadian companies, including Mabarex (http://www.mabarex.com/en/index.php). However due to the problems experienced with the technology the licence agreements were terminated. Several years ago Phosphex technology was licenced for the US market to Lombardo Associates Inc., who have exclusive rights to sell Phosphex systems. Nonetheless, at the Lombardo Associates website, there is no reference to Phosphex technology (Lombardo Associates 2018).

Due to the concerns about the potential metals leachate from the filtration media and the lack of regulatory requirements for P removal from OSS in the USA less than a dozen systems were installed in the past 15 years.

### 3.4.2   PhosRID™

Upon obtaining the licence for Phosphex technology, Lombardo Associates Inc. found that it did not meet required performance. They thus modified the media by adding larger quantities of limestone in the mix of materials and named the new process PhosRID technology (B. Eichinger, personal communication). PhosRID systems are housed in concrete tanks, and similarly to Phosphex effluents with highly elevated pH; they are offered at the price of 12–15 000 US$/single household (B. Eichinger, personal communication). Material replenishment via injection is recommended every 7–10 years, however, to date there is no evidence whether this method would work in practice or not (Lombardo Associates Inc. 2018).

A PhosRid system implemented at the Massachusetts Septic System Test Center 10 years ago is described in a PhisRid system brochure presented at Lombardo Associates website (Lombardo Associates 2018).

### 3.4.3   PhosphoReduc™

The PhosphoReduc system is a 'closed loop' gravity fed passive filtration system for P harvesting/removal, re-use, and recovery, and as such also the enabler of the circular economy. It was co-invented by Drizo and Picard, who hold the rights for the PhosphoReduc technology in the US (Drizo and Picard 2014). The technology combines pre-selected filtration media (modified SSA) and unique design to ensure P removal, harvesting, and reuse from any point or nonpoint pollution sources (NPPS).

#### 3.4.3.1   Filtration Media

SSA are an industrial recyclable by-product from the steel manufacturing industry (Proctor et al. 2000). Steel is the grounding component of the world economy with 1.7 billion tons produced annually in over 65 countries (WorldSteel Association 2018). The steelmaking process involves melting iron ores and scrap in large furnaces, along with fluxing agents and the molten impurities of the steel scrap. After the molten material is cooled and solidified, the metallic component is removed (fed back into the steel mill) and the non-metallic component is crushed to form slag aggregates of various sizes (Proctor et al. 2000; Drizo et al. 2002). They are stockpiled on steel mills across the world in millions of tons with the major use in roads building and other construction industry, and to a smaller extent in agriculture (instead of lime). Given that the SSA aggregates are rich in Ca and Fe oxides (on average 30% and 25% respectively), their vast geographic availability and low cost (due to waste legislation that classifies SSA as a waste product), they have been extensively investigated for P removal from synthetic, man-made wastewaters in laboratory experiments across the world (Yamada et al. 1986; Johansson 1997; Baker et al. 1998; Johansson 1999a, b; Forget 2001; Drizo et al. 2002; Kostura et al. 2005; Chazarenc et al. 2008; Drizo et al. 2008a, b; Johansson-Westholm et al. 2010; Lee et al. 2010). Similar to other filter materials, research on using SSA to treat real wastewater had been much less extensive. The SSA research on phosphorus removal from wastewaters has been thoroughly reviewed by Chazarenc et al. (2008), Johansson-Westholm et al. (2010), Johansson-Westholm (2010), Vohla et al. (2011), and Barca (2012).

Blast furnace (BF), iron melter slag (IMS), and electric arc furnace (EAF) are the three types of steel slag materials that were most frequently investigated. Each of these materials has specific mineralogical and physico-chemical properties that are dependent on the steelmaking processes used. It is thus of utmost importance to investigate materials' chemical composition prior to any use in the field (Drizo 2012). The BF steel slag originates from integrated steel mills where iron ore is melted utilizing BFs. The IMS is produced from iron sand found only in Glenbrook, Auckland, New Zealand. The EAF steel slag is produced by melting scrap steel in an industry known as 'mini-mills' (Proctor et al. 2000; Drizo 2012; Drizo et al. 2008b).

### 3.4.3.2 Technology Background

The technology was developed as an outcome of a decade (1999–2009) of research by Drizo and co-workers on the use of SSA for P removal from wastewaters (Drizo et al. 2002, 2006, 2008b; Weber et al. 2007; Bird and Drizo 2009, 2012). They were amongst the first researchers who conducted series of field-scale investigations on the potential of SSA for P removal from a variety of wastewater effluents (dairy, farmyard runoff, surface and subsurface agricultural drainage, urban stormwater runoff, industrial sites runoff, and sewage). Between 2004 and 2008 they established over 15 different long-term pilot and medium-scale experiments and investigated a number of operating parameters known to affect filters' field performances. This extensive research resulted in the development of operational parameters for six different classes of technologies for phosphorus, SS and pathogens reduction and phosphorus harvesting, recycling and re-use from any point or NPPS (Drizo 2012; WSSI 2018). The establishment of a small business PhosphoReduc in 2007 enabled further media investigation, technology development, and performance assessments for treatment of agricultural and urban wastewaters and runoff, and municipal and residential effluents across four continents (North America, South America, Asia, and Europe). These projects showed on average 90% phosphorus, 95% pathogens, and 90% SS reduction (WSSI 2018). In addition, Drizo and Picard (2014) co-invented a pH reducing unit as an integral component of a PhosphoReduc system which sets this technology ahead of all other technologies based on alkaline media (e.g. Polonite, Filtralite, Phosphex, and PhosRid). A range of PhosphoReduc projects and applications can be seen on the Water and Soil Solutions website (WSSI 2018).

### 3.4.3.3 PhosphoReduc System for Onsite Residential Wastewater Treatment ('PR-G-1000')

This is a simple, gravity-fed passive filtration system that can be placed between the septic tank and a soil adsorption system or leachate field or as an 'add-on' component to any Nitrogen removal system (WSSI 2018). P is harvested/removed from the waste stream via precipitation with Ca oxides in SSA media when designed at specific hydraulic residence times (HRT) of 18–24 hours. Due to hydrolysis of CaO and subsequent dissociation in solution, SSA also generates highly elevated pH (12–14) at the effluent. Therefore, Drizo and Picard (2014) co-invented a pH reducing unit as an integral component of a system which sets this technology ahead of all other technologies based on alkaline media presented earlier in the chapter. The system is housed in a standard septic tank (volume $4.4\,m^3$), resulting in a small footprint (surface area $5.5\,m^2$). The tank is divided into two compartments by a baffle wall: in the first compartment, P is harvested/removed from the waste stream whilst in a second compartment pH is reduced and adjusted via a unique mix of organic media (Drizo and Picard 2014).

Although PhosphoReduc has obtained several permits for the system use as an alternative technology for P reduction from onsite residential wastewater treatment systems (WSSI 2018), due to the absence of regulatory requirements to date only five systems have been implemented commercially. Of these four in the USA (North Carolina, 2012; Michigan, 2015, 2016; and New York State, 2015) and one in Europe (Ireland, 2014) (Figure 3.5). Independent water quality monitoring of the system implemented in Ireland showed 80% reduction in total Phosphorus (TP), 96% reduction in TSS, 85% reduction in organic matter (BOD), 65% reduction in TN and 74% reduction in ammonia ($NH_4^+$).

The cost of the system is ~US\$ 7000 (6174 €)/single household with estimated US\$ 3500 (3000 €) for the media replacement every 5–7 years, making it the cheapest passive filtration technology (US\$ 17 500 or 15 435 € over 25 years period) currently available on the onsite wastewater treatment market (Table 3.2).

(a)

(b)                                          (c)

**Figure 3.5** PhosphoReduc System implemented in (a) Ireland, (b) North Carolina, and (c) Michigan. Source: Water and Soils Solutions International (2018). www.phosphoreduc.com

**Table 3.2** Passive filtration systems comparison.

|  | Polonite P traps | Filtralite P | Phosphex | PhosRid | PhosphoReduc |
|---|---|---|---|---|---|
| P reduction | Over 90% | Over 90% | Over 90% | Over 90% | Over 90% |
| P reuse potential | Yes | Yes | Not reported | Not reported | Yes |
| pH | Over 12 | Over 12 | Over 12 | Over 12 | 9 |
| Media replacement frequency (years) | 2 | 15 | 7–10 | 7–10 | 5–7 |
| Total Cost over 25 years period (€)[a] | 32 318 | 20 000 | 20–23 000 | 20–23 000 | 15 435 |
| Initial capital cost (€) | 6314 | 15 000 | 10–13 000 | 10–13 000 | 6174 |
| Number of systems implemented | 5000 | ~ 100 | < 10 | < 10 | < 10 |
| Geographic region | Sweden, Norway, Finland | Norway | Canada, USA | USA | USA, Ireland, Turkey |

[a] Conversion to euros on 10 November 2018.

### 3.4.4 Commercial Products Based on Electrochemical Processes

The electric coagulation (EC) process has been used for the treatment of industrial wastewaters for most of the twentieth century with limited success and popularity (Mollah et al. 2001). In 2013, two new products based on the EC process were advertised on the North American market for onsite residential wastewaters treatment: Fuji Clean P removal process model CRX (offered by Fuji Clean Company, Japan) and DpEC Self-Cleaning P Removal Unit (offered by PremierTech, Quebec, Canada). In its simplest form, the EC process consists of a reactor made up of an electrolytic cell consisting of one anode and one cathode. A detailed description of a process and its advantages and disadvantages is provided by Mollah et al. (2001).

The Fuji Clean P removal system for residential effluents treatment ('Phosphorus Removal Device') requires pre-treatment via an aerobic and anaerobic filtration chamber (Fuji Clean 2018). Phosphorus is removed by passing a direct current through a pair of iron electrodes immersed in water during which divalent iron ion ($Fe_2^+$) is extracted from the anodal electrode and oxidized with dissolved oxygen to the trivalent iron ion ($Fe_3^+$). P is precipitated from wastewater as the insoluble iron phosphate ($FePO_4$) and carried back to the head of the system via airlift pumps where is eventually removed as sludge. A detailed process description and system components are described at Fuji Clean website (Fuji Clean 2018). However, the information on the systems installation, performance, costs, or the potential for P recovery is not available.

Similar to the Fuji Clean P removal device CRX, PremierTech DpEC Self-Cleaning P Removal Unit contains an EC unit that receives pre-treated effluent from a primary reactor system. The main difference between the two systems is that the later one uses aluminium electrodes (PremierTech Aqua 2018a). According to several brochures and fact sheets generated by Premiertech, testing of the DpEC Self-Cleaning P Removal Unit for use in the Quebec market showed that P concentrations as low as 0.4 mg/l are achievable (PremierTech Aqua 2018b). However, information on the prior research, systems field implementation, or the costs is not provided (PremierTech Aqua 2018a, 2018b).

Whilst both the Fuji Clean P removal device CRX and PremierTech DpEC Self-Cleaning P Removal Unit products claim significant P reductions there is very limited information on the actual testing and field performance and the products' installation, operation, and maintenance costs.

Electrodes are known to be prone to corrosion and need to be replaced regularly. An impermeable oxide film may form on the cathode leading to the loss of efficiency (Mollah et al. 2001). The EC products may require complicated electrical authentication of facilities, over-consumption of electric power and may pose potential risks to the homeowners such as electric shock injury accidents (Hong et al. 2013).

## 3.5   Agricultural Phosphorus Pollution and Mitigation Measures and Strategies

Despite over 40 years of concerted efforts and considerable financial investments in implementation of conservation control interventions and management strategies, agricultural P pollution and its detrimental impacts on water quality remains a major environmental issue (e.g. Bomans et al. 2005; Schoumans et al. 2011; Schoenberger et al. 2014 [Europe]; Jarvie et al. 2013; Withers et al. 2014b [UK]; Kleinman et al. 2011; Kleinman et al. 2015 [USA]; Sharpley et al. 2015 [North America and Europe]).

The following sections will discuss P input from agricultural production, the origin, and history of agricultural 'best management' practices (BMPs, term used in the USA), the costs of BMPs and 'good agricultural practices' (GAP, term used in Europe) implementation, and methods and challenges in assessing their cost-effectiveness. Given the large body of literature that provides description and fact sheets on dozens of BMPs and GAPs for agricultural P pollution mitigation (see Table 3.5) the focus of this section will be on those most commonly implemented, treatment performances and a few novel methods and their treatment performances. In addition, challenges and barriers to technological innovation research and development, deployment and adoption on farms will also be described.

### 3.5.1   Phosphorus Input from Agricultural Production

#### 3.5.1.1   Livestock

Agricultural production (livestock, food and non-food crop commodities) has long been recognized as the greatest contributor to P pollution and subsequent eutrophication (Duda and Finan 1983; Edwards and Daniel 1992; Sharpley et al. 1994; Sharpley et al. 2011; Kleinman et al. 2011; Jarvie et al. 2013; Withers et al. 2014b; Kleinman et al. 2015; Sharpley et al. 2015). The total agricultural production has tripled during the past 50 years and has undergone a dramatic change in management practices (Wik et al. 2008). Today, the livestock sector is associated with the development of industrial and intensive production systems where a large number of animals are concentrated in a relatively small area (e.g. Concentrated Animal Farm Operations, CAFOs) posing a continuous pollution threat to adjacent waters. In the USA, for example, beef, dairy, pig, and poultry numbers have increased 10–30% between 1990 and 2004, whilst the number of farms has decreased 40–70% (Sharpley et al. 2011). Similar conglomeration has occurred in Europe and Australasia, driven by incessant population growth and demand for food and animal products profitability. According to the FAO (FAO 2015a), pig production accounts for an estimated 90% of P input into the South China Sea from the Pearl River basin in Guangdong Province, the Chao Phrya River basin in Thailand and the Red River and Dong-Nai

River basins in Viet Nam. Pig and poultry meat and meat products are the most popular meat consumed in China, which is also the world's largest livestock producer and consumer and has 400 million cattle, sheep, and goats (Agrifood Asia 2018). The total demand for animal products in developing countries is expected to more than double by 2030 (FAO 2015b). The global cattle population reached 1 billion in 2012 (Statista 2018b) and is projected to reach 1.9 billion by 2030. The number of pigs in 2011 was 968 million, projected to reach 1.06 billion by 2030; the number of sheep and goats will reach 2.3 billion. The number of poultry in 2011 was 19 billion and is estimated to reach 24 billion by 2030 (*The Economist* 2011; FAO 2015b). Livestock excreta contains considerable amounts of nutrients, veterinary drug residues, heavy metals, and pathogens (Box 3.2).

Apart from P concentrated in excreta and manure, livestock operations discharge considerable quantities of P via stormwater runoff from open manure pits and storage facilities, animal farmyards and buildings' roofs, feedlots, silage and composting piles, dairy milking parlours, slaughterhouses, and other heavy animal use areas. Of all the agricultural discharges, silage leachate represents one of the potentially most contaminated and harmful wastes generated on a farm. Due to its high BOD, nutrient-rich composition (up to 600 mg/l of phosphorus) and low pH, silage leachate is approximately 200 times stronger than raw domestic sewage and 40 times stronger than dairy shed waste (Bloxham 1992; USDA NRCS 1995). Feed bunks runoff also contains high concentrations of P; for example, Drizo et al. (2008a) measured dissolved P concentrations of 100–130 mg/l, averaging 41.4 mg/l over the period of 225 days at the University of Vermont dairy farm located at Burlington, Vermont, USA. Additional sampling and measurements during a two-year period (2009–2010) at the same farm revealed that dairy parlour wash water and farmyard feedlot runoff dissolved P concentrations (44.1 mg/l and 32.6 mg/l respectively) exceeded those found in manure runoff (16.7 mg/l). These heavy animal use areas represent the critical pollution source areas from which phosphorus ought to be reduced. In addition, the potential for P harvesting, recycling, and re-use should also be explored.

---

Box 3.2  Phosphorus Content in Livestock Excreta and Manure

A dairy cow excretes on average 6.1 (less productive areas) to 16 kg P/year (high productive areas), a sow 4.7–8 kg P/year, growing pig 2.3–2.5 kg P/year, layer hen 0.14–0.22 kg P/year and broiler 0.1 kg P/year (De Wit et al. 1997). A detailed standard containing complex equations for estimating livestock and poultry manure production and characteristics based on typical diets and animal performances in the USA was developed by the American Society of Agricultural Engineers (2005). Sun and Wu (2013) used unit load and export coefficient modelling to estimate pollutant loads from livestock and poultry raising in China from year 2000 to 2010. The results revealed vast P loading of 6 million metric tons/year.

### 3.5.2 Crop Production

Monfreda et al. (2008) compiled a detailed database of global land use practices and geographic distribution of crop areas, yields, physiological types, and net primary production for the year 2000. They estimated that cropland land use covered ~15 million km$^2$ of the planet land surface area. According to the 2015 World Bank database, agricultural land comprising of land area that is arable, under permanent crops, and under permanent pastures covered 48.6 million km$^2$ of the planet land surface area (World Bank 2018a).

Agricultural producers and farmers apply millions of tons of chemical fertilizers and manure to improve crop yields. In addition, there is a vast variety in the rates of application and many fields may receive a mix of manure/fertilizer types in several applications over a single growing season (Kinley et al. 2007). The type of manure applied has an effect on P losses as poultry manure may contain 15 and 10 times more P than dairy and swine manure. However, grassland fields sometimes receive 10 times more dairy manure volume than fields receiving poultry or swine manure (Kinley et al. 2007).

The global use of fertilizers increased 19-fold in the last century, from about 873 million tonnes of P in 1913 to about 16 591 million tonnes of P in the late 1980s (Hart et al. 2004). The World Bank 2015 data (World Bank 2015) provides comprehensive data on the global chemical fertilizer (nitrogenous, potash, and phosphate fertilizers) consumption per country, measured as the quantity of plant nutrients used per unit of arable land (excluding plant and animal manures). A 2018 World Bank database provides important data on land surface area, percent of arable land, and annual fertilizer consumption (kg/ha) (World Bank 2018a, b, c, d). Data for the five largest countries (by land surface area) are presented in Table 3.3. Although the USA and China occupied third and fourth place in the land surface area, they have the largest percentage of the arable land. China has by far the highest fertilizer consumption of 506.1 kg/ha resulting in an immense total

**Table 3.3** Annual chemical fertilizer consumption per country.

|  | Land surface area (millions ha)[a] | Percent of arable land[b] (%) | Fertilizer consumption[c] (kg/ha) | Total Fertilizer consumption (tons) |
|---|---|---|---|---|
| Russian Federation | 17 098.0 | 7.5 | 16.5 | 21 158.8 |
| Canada | 9984.7 | 4.8 | 91.6 | 43 900.7 |
| USA | 9831.5 | 16.6 | 137.0 | 223 588.0 |
| China | 9562.9 | 12.7 | 506.1 | 614 652.5 |
| Brazil | 8515.8 | 9.6 | 163.7 | 133 827.5 |

[a] Source: World Bank (2018b).
[b] World Bank (2018c).
[c] World Bank (2018d).

annual fertilizer consumption of 614 652 tons; the USA has the second highest consumption, followed by Brazil (Table 3.3).

Phosphorus freshly applied in chemical fertilizers and manure, or accumulated in soils from decades of regular fertilizer applications is frequently lost from the soil through leaching, subsurface and surface runoff creating numerous diffuse pollution sources (e.g. Kleinman et al. 2011; Sharpley et al. 2011). Additionally, in many instances, agricultural tile drains serve as conduits of phosphorus rapid transport from the fields to adjacent waters (Kinley et al. 2007; King et al. 2014).

### 3.5.3   Pasture, Rangeland, and Grazing Operations

Grasslands, sown pasture, and rangeland, are amongst the largest ecosystems in the world and contribute to the livelihoods of more than 800 million people. They are a source of goods and services such as food and forage, energy, and wildlife habitat, and also provide carbon and water storage and watershed protection for many major river systems. The FAO estimated the world area of pasture and fodder crops at 3.5 billion ha (35 000 000 km$^2$) in 2000, representing 26% of the world land area and 70% of the world agricultural area (FAO 2015c). Overgrazing along pasture and rangeland streams makes an additional contribution to phosphorus loading via direct deposition of faeces in the proximity to the water, precipitation runoff, and stream erosion (e.g. Haan et al. 2006; Russell et al. 2006; Alexander et al. 2008; Schwarte et al. 2011).

### 3.5.4   Agricultural BMPs – Origin and Brief History

The Agricultural BMPs (North America) or 'GAP or GPs' or 'Best environmental management practices (BEMPs)' (Europe) are broadly defined as methods and/or practices designed to reduce or prevent soil and water pollution without affecting farm productivity (Logan 1993; Hilliard et al. 2002; Sharpley et al. 2006; OMAFRA 2015). They were developed in 1950s as the conservation measure to combat soil erosion and were implemented as soil remediation practices for two decades prior to the first awareness and recognition that non-point source (NPS) phosphorus from soil runoff and erosion, fertilizer, and livestock runoff results in eutrophication of water bodies in 1970s (Logan 1993). In the USA, development, and implementation of conservation BMPs to protect water resources became a requirement in the 1977 amendment of the Clean Water Act (US EPA 2015). By the early 1980s the National and State Planning Agencies started to develop remedial plans to deal with agricultural NPS P pollution, based on voluntary adoption by farmers (Logan 1993). The same BMPs concept based on voluntary adoption by farmers remained a dominant tool in agricultural soil and water pollution mitigation for the following 40 years.

Logan (1990, 1993) was amongst the first researchers who proposed BMPs categorization according to the environmental objective, target pollutant type, the environmental medium impacted, and management approach as (i) Structural controls; (ii) Source Controls; (iii) Land Management Practices; and (iv) Pest management Practices (Table 3.4). He also discussed the BMPs effectiveness in the management and mitigation of some of the key pollution sources, e.g. livestock management and P fertility management. He underlined the fact that most mitigation efforts were focused on cost share for storage facilities neglecting to address a persistent problem of the inadequate land base for agronomic utilization of livestock waste nutrients. Last but not least, he signalled the fact that a large number of agricultural soils in the USA have available P levels in the excessive range and a need for better P fertility management (Logan 1993). Today, after decades of continuous manure and fertilizer application, excessive P levels in soils are often referred to as the 'Legacy P' and remain one of the major challenges in mitigating agricultural P pollution from NPS (Kleinman et al. 2011; Jarvie et al. 2013; Withers et al. 2014b; Sharpley et al. 2015).

**Table 3.4** Agricultural Best Management Practices (BMP) classification according to Logan (1990).

*Structural Controls*

| | |
|---|---|
| Purpose | Designed primarily to modify pollutant transport in water by reducing water use, rerouting or retaining water and included terraces, grassed waterways, buffer strips, tile drains, irrigation systems, livestock waste storage facilities, and sediment detention basins. |
| Advantages | Effective for control of sediment and sediment-associated pollutants in surface runoff, and may offer benefits for groundwater protection. |
| Disadvantages | High capital costs and their implementation on farms required cost-sharing incentives. In addition, they generally require long-term maintenance |

*Source Controls*

| | |
|---|---|
| Purpose | They primarily affect the source of a potential contaminant by increasing use efficiency and include management practices for fertilizer, pesticide, livestock waste application, and soil fertility management. |
| Advantages | The most effective practices and easiest to regulate. Nutrient application rates can be restricted on the basis of the potential for surface water or groundwater contamination. |
| Disadvantages | Problems with the imposition of nutrient application limits included unpredictable soil nitrogen and phosphorus transformations, losses that could reduce crop yields, and difficulty in administering and regulating nutrients use restrictions. |

*Land Management (Cultural) Practices*

| | |
|---|---|
| Purpose | Management practices that manipulate the soil system to minimize pollutant losses in surface water or groundwater. These include (i) timing and placement of nutrients to achieve maximum effect and minimum carryover; (ii) application methods for livestock waste to reduce runoff; (iii) irrigation scheduling to minimize water use and excessive leaching, and (iv) conservation tillage for runoff and erosion control. |
| Advantages | Showed some effectiveness in individual studies. |
| Disadvantages | Specific criteria difficult to develop, and require significant, extended educational efforts to attain large-scale impact. Monitoring of adoption rates is difficult on a large scale. |

### 3.5.5   BMPs and GAPs Guidelines and User Manuals

Since the early 1990s, the research on agricultural BMPs, including those for P management has increased across the globe. The simplest web search engines will generate a myriad of agricultural BMPs User Manuals and Guidelines for almost every country/state/province/region in the world. Some of the examples include: Browning et al. 1996 and Cuttle et al. 2007 (UK); Lam et al. 2011 (Germany); Sims et al. 2005 (Europe and North America); Lietman et al. 1996; Bracmort et al. 2004; Sharpley et al. 2004, 2006; Coxe and Hedrich 2007 (USA); Hilliard et al. 2002; Agriculture and Agri-Food Canada 2004; Tamini 2009 (Canada); European Commission 2015b (Europe); Premier and Ledger 2006 (Australia and Southeast Asia); Collins et al. 2007 (New Zealand); FAO 2013; Xie et al. 2015 (China); and Teenstra et al. 2014 (Global).

In Europe, Bomans et al. (2005) generated a comprehensive 283 pages report for the European Commission titled 'Addressing phosphorus-related problems in farm practice' (Table 3.5). The report reviewed the role and use of P in the agricultural sector of the European Member States, P legislation within each member state, and legal and practical measures that can be taken to reduce the losses of P from agricultural activities to the aquatic environment. It also described Good Agricultural and Environmental Condition (GAEC) Practices and their cross-compliance requirements beyond EU environmental legislation. As the focus in the EU has been on P use in agricultural application and thus the most commonly implemented BMPs were: the reduction of nutrient application, the modification of culti-vation techniques, fertilizers and pesticides handling and soil cultivation

**Table 3.5** List of Reports describing agricultural best management practices (BMPs) and good agricultural practices (GAPs) to address and mitigate agricultural P pollution 2005–2018.

| Year | Title | Pages | Author |
|------|-------|-------|--------|
| 2005 | Addressing phosphorus related problems in farm practice | 283 | Bowmans et al. (2005) |
| 2005 | SERA-17 Phosphorus BMP Factsheets | | SERA 17 (2018) |
| 2006 | BMP To Minimize Agricultural Phosphorus Impacts on Water Quality | 52 | Sharpley et al. (2006) |
| 2011 | Mitigation options for reducing nutrient emissions from agriculture. A study amongst European member states of Cost action 869. | 147 | Schoumans et al. (2011) |
| 2012 | BMP Policy Tool Box Presentation | 18 | UNEP (2012) |
| 2014 | Development of the EMAS Sectoral Reference Documents on Best Environmental Management Practice (BEMP). Learning from frontrunners promoting best practice. | 26 | Schoenberger et al. (2014) |
| 2018 | EU Database of Best Practices | | Living Water Exchange (2018) |
| 2018 | BEMP for the agriculture sector – crop and animal production | 628 | European Commission (2018a) |

techniques to prevent soil erosion. Thus most measures were aimed at reducing the influx of nutrients to ground and surface waters.

In the same year (2005) the European Fifth Framework Programme funded another long-term research project to investigate Mitigation options for nutrient reduction in surface water and groundwater, which involved participants from 30 European countries (Box 3.3). The major findings from this research were published in a ~150 pages report titled 'Mitigation options for reducing nutrient emissions from agriculture' (Schoumans et al. 2011).

---

Box 3.3  Nutrient Reduction Cooperation in Science and Technology (COST) Project

In order to reach targets of the WFD (Chapter 2, Section 2.2), in early 2005, an ambitious proposal was launched to the Fifth European Framework Programme for a new COST action under the Agriculture, Biotechnology and Food Sciences Programme. It was approved in 2006 as the COST action 869 'Mitigation options for nutrient reduction in surface water and groundwater'. Thirty countries participated in the project: Austria, Belgium, Bulgaria, Czech Republic, Denmark, Estonia, Finland, France, Germany, Greece, Hungary, Ireland, Israel, Italy, Latvia, Lithuania, Luxembourg, Netherlands, New Zealand, Norway, Poland, Portugal, Romania, Slovakia, Slovenia, Spain, Sweden, Switzerland, Turkey, and the United Kingdom. As a result of this comprehensive research an extensive report has been generated containing agricultural pollution mitigation options, factsheets. and costs (http://www.cost869.alterra.nl).

---

### 3.5.6   Today Europe Remains Very Far from Curtailing Phosphorus Pollution from Agriculture

In parallel with the initiation of the European COST project, in the USA, the Southern Extension and Research Activity Information Exchange Group was formed (SERA 17) with a mission to minimize P losses from agriculture through innovative solutions. It consists of national and international research scientists, policy makers, extension personnel, educators, and regulatory communities (SERA 17 2018).

According to the United Nations Environmental Programme (2012), the current global inventory contains 290 best nutrient management practices for agriculture pollution mitigation. The information on these practices was compiled from over 55 organizations (including the Global Environment Facility (GEF), the United Nations Development Programme (UNDP), the USA Natural Resources Conservation Service (NRCS), EU and the World Bank) and over 55 different countries in North and South America, Europe, Africa, and Asia. In addition, Schoenberger et al. (2014) recently generated the EU Joint Research Committee (JRC) Scientific and Policy report, which represents a compilation of the Eco-Management and Audit Scheme (EMAS) Sectoral Reference Documents on BEMP (Table 3.5).

The investment in gathering information on 'best or good' practices for agricultural management and environmental protection continues to support production of comprehensive, lengthy reports (e.g. European Commission 2018a) and projects. In 2016, the European Union European Research Development Fund Northern Periphery and Arctic Programme 2014–2021 awarded 1.7 million euro to the University of Savonia, Finland and 23 European partners (Finland, Sweden, Iceland, Faroe Islands, Ireland, and Scotland) for a project on BMPs for Agricultural and Minerals Extraction Runoff treatment. Some of the main objectives of this project are to build and create the inventory of current management practices and technologies for agricultural and minerals extraction runoff treatment, climate change adaptation, and disaster preparedness in remote and sparsely populated communities of the Northern Arctic Periphery Region (http://www.water-pro.eu).

### 3.5.7   The Costs of Agricultural Management Practices' (AMPs) Implementation

The OECD study reported that the range of BMPs deployed at the local, catchment, regional, national, and international scales across an array of different governmental agencies cost taxpayers billions of dollars annually (OECD 2012). The example of some of the highest agro-environmental monetary payments for selected countries is presented in Table 3.6.

Governmental expenditures typically include (i) the financial agro-environmental payments provided directly to agricultural producers as a compensation for a loss of income for adopting sustainable agricultural conservation management practices and (ii) expenditures on various technical assistance for BMP implementation (OECD 2012; Shortle et al. 2013). In the EU27 these payments are fully financed by the EU, and

**Table 3.6** Agro-environmental monetary payments to farmers and other landholders over two year period (2007–2009).

|  | Annual Agri-environmental monetary payments[a] | Percentage of the Producer Support Estimate (PSE)[b] |
|---|---|---|
| EU27 | € 6000 million | 7 |
| USA | € 4400 million ($ 5 billion USD) | 16 |
| Australia | € 286 million (450 million AUD) | 18 |
| Switzerland | € 211.5 million (240 million CHF) | 4 |
| Norway | € 97.6 million (950 million NOK) | 5 |

[a] Currency conversion rates calculated in November 2018.
[b] The OECD defines the PSE as an indicator of the annual monetary value of gross transfers from consumers and taxpayers to support agricultural producers, measured at farm gate level, arising from policy measures, regardless of their nature, objectives, or impacts on farm production or income.

account for 70% of the Common Agricultural Policy budget (European Commission 2015c).

In the USA, the investment in agriculture is determined by the US Congress through farm bills (Chapter 2). The USDA Farm Service Agency (FSA) administers the Conservation Reserve Program (CRP) which they consider as one the most successful voluntary conservation initiatives in US history (USDA 2018). It is a cost-share and rental payment programme which provides hundreds of millions of dollars in federal funds for the implementation of BMPs recommended by the USDA Forest Service and the USDA NRCS (USDA NRCS 2018). However, the Program does not provide any background or information regarding the rules or protocols for novel solutions acceptance as a conservation national practice standard and a practice. The USDA NRCS website only states that 'one must have the conservation practice standard developed by the state in which the person is working to ensure that you meet all state and local criteria, which may be more restrictive than national criteria'. And that National Conservation Practice standards should not be used to plan, design, or install a conservation practice. Details of this arduous and long process for the State of Vermont are described in Chapter 2, Case study 3. It appears that once the novel solution is accepted as an interim conservation standard and/or practice in one State, its acceptance in other States becomes a relatively straightforward process. However, as described in Case Study 3, despite seven-plus years invested in research, development, verification, and negotiating with the relevant USDA NRCS and agricultural agency staff, the researcher/inventor does not receive any kind of recognition or a credit for the new conservation practice. The costs of novel practice design and implementation are determined by the individual USDA NRCS offices as it is being offered by the USDA NRCS staff.

In most cases, the USDA NRCS Conservation Practices Program does not provide any funds for the assessment of BMPs performance in reducing P (and other) pollutants. Instead, funding for water quality protection has often been allocated for implementation of the BMPs for which there has been no or very little evidence on the measurable effectiveness in P reduction, for example, conservation buffer strips. Consequently, after years of investments in BMPs' implementation, their ability to mitigate P pollution remains largely unknown (Section 3.5.10). The USDA recently recognized the need for BMPs and long-term catchment monitoring and new conservation strategies that provide cost-share funds to farmers have begun to require that 10% of the funds are allocated to the monitoring of the BMPs' effectiveness. In addition, the USDA established a standard for monitoring of the edge-of-field surface runoff water quality, which became a pre-requisite for farmers' eligibility to the financial assistance programme (Sharpley et al. 2015).

In addition to the USDA, the US Environmental Protection Agency (USEPA) supports programmes to reduce the negative impacts of runoff from agricultural, urban, and industrialized areas under the Clean Water Act Section 319 Nonpoint Source National Monitoring Program and wetland protection programmes. However, these programmes are also focused on BMPs' implementation with very limited funds for the evaluation of their effectiveness.

The costs of BMPs' implementation in the USA is well illustrated by the example of Chesapeake Bay watershed which has the largest estuary (167 000 km$^2$), drains in six states and is severely affected by eutrophication and HABs. Acknowledging the lack in water quality improvement despite decades of investments, in 2009, President Obama issued Executive Order 13508 for Chesapeake Bay Watershed Protection and Restoration (The White House 2009). To fulfil the Executive Order, the USEPA developed Total Maximum Daily Load (TMDL) for the Bay, the Watershed Implementations Plans (WIPs) and Nutrient trading programmes and tools for the states within the Watershed to reduce nutrients loading (Chapter 2). However, it has been estimated that the cost of implementing the required BMPs for reducing agriculture's nutrient discharges between 2011 and 2025 would be $3.6 billion (in 2010 dollars), whilst the costs associated with full implementation of all WIP BMPs from 2025 onwards (2025 annual costs) would be $900 million/year (Shortle et al. 2013). The researchers suggested that better BMP selection and spatial implementation targeting could result in significant cost savings. However, whether greater cost savings can be achieved by a targeted, plan-based method or the optimization of a performance-based method is a question of debate worldwide (e.g. Srivastava et al. 2002; Veith et al. 2003; Panagopoulos et al. 2011; Talberth et al. 2015).

In November 2003, FAO published a Summary analysis of codes, guidelines, and standards related to Good Agricultural Practices, GAP (FAO 2003). The report also described developments in public-funded standards, guidelines, and incentives schemes on GAP as well as the development of social and environmental voluntary standards and certification programmes. A 2017 European Commission report summarizes direct payments to farmers in the EU for the period 2015–2020. It states that over €41 billion a year has been spent for direct payments to farmers during the 2015–2020 period (European Commission 2015c).

The cost-effectiveness of BMPs has been a subject of numerous scientific papers and yet it remains unknown. The *Land and Policy Journal* 2010 (Volume 27, issue 1) published a special issue on Soil and Water Conservation Measures in Europe, where issues regarding the determination of BMPs cost-effectiveness and adoption were described in 12 scientific papers. In 2011, the European Cooperation in Science and Technology (e-COST) programme concluded a five-year scientific evaluation of the suitability and cost-effectiveness of different options for reducing nutrient loss to surface and ground waters at the river basin scale. Findings from this long-term research resulted in 14 papers published in a special issue of the *Journal of Environmental Quality* (issue 2, 2012). In 2013 the Seventh International Phosphorus Workshop (IPW7) was held in Sweden with over 150 delegates involved in roundtable discussions on the management of agricultural P to minimize impacts on water quality. These discussions were summarized in a series of papers published in a special issue of *AMBIO* journal in 2015.

### 3.5.8 Methods for Assessing BMP's Cost-Effectiveness in Mitigating Agricultural P Pollution

Cherry et al. (2008) conducted a comprehensive review of methods for assessing the effectiveness of actions to mitigate agricultural nutrient pollution categorizing them into (i) measurements, (ii) nutrient budgets, (iii) risk assessment, (iv) modelling, and (v) an integrated approach to assessment. Acknowledging the complexity of an array of environmental processes involved, lack of understanding of nutrient dynamics (particularly in-stream processes and ecological responses), and variations in weather and its effects between years, the authors underlined the necessity for installation of automated in situ sampling and analytical equipment to facilitate high-frequency sampling. In addition, they stated that long-term series of measurements of nutrient concentrations and mass loads in water bodies provide the best analysis and assessment of mitigation success as they describe the actual change in chemical composition and water quality following the implementation of mitigation.

The need to reduce costs of BMPs implementation resulted in a paradigm shift in agricultural P management strategies from unilateral recommendation of conservation measures to address P loss across a watershed towards specific targeting of particular BMPs on critical source areas (CSA) within a watershed anticipated to contribute most heavily to NPS pollution (Veith et al. 2003; Sharpley et al. 2011). This approach was based on the watershed research findings which showed that the majority (~ 80%) of the P loss originates from only a small proportion (~20%) of the watershed. Consequently, targeting was developed as a 'plan or performance-based method' and was based on physical characteristics or cropping practices, focusing on field parameters such as slope, soil type, proximity to stream, crop, and tillage practices (Veith et al. 2003).

Taking into account the complexity and site-specific nature of P losses from agriculture, Kleinman et al. (2011) discussed key opportunities, challenges, practices, and strategies to control agricultural P pollution and enhance water quality. They distinguished acute, temporary (recently applied P as manure/fertilizer) and chronic sources of P ('legacy' P in soils accumulated over the years) and proposed several principles for progress to achieve better water quality protection. These included (i) adjustments in agricultural P management to better address the role of hydrology on P transfers (e.g. mobilization, transport and delivery from field to water body); (ii) taking into consideration the potential for subsurface transport of P (tile drainage and tillage management); (iii) P legacy (underlying the need for novel practices to confront P legacy); and (iv) the need to control P in pasture-based systems. Moreover, they acknowledged the potential of novel practices to intercept surface and subsurface farm flows (drainage lines and ditches) and adjacent to areas of acute P accumulation (e.g., farmyards, subsoils). Recognizing that CSA should be targeted for cost-effective remedial management, they underlined that identifying these areas is a difficult task, even in areas with an abundance of agronomic and physiographic data.

Sharpley et al. (2011) corroborated principles outlined by Kleinman et al. (2011) and suggested that other long-term factors, such are farm- and watershed-scale P imbalances and vertical stratification of P in soils must be taken into consideration in the selection of the BMPs locations. In addition, they reviewed Phosphorus Index, a site assessment tool which was designed and proposed 20 years ago with the aim to identify and rank CSA of P loss based on site-specific source factors (soil P, rate, method, timing, and type of P applied) and transport factors (runoff, erosion, and proximity to streams) and has been adopted in as many as 47 US States. Furthermore, they pointed out that in some cases BMPs implemented to reduce one form of P loss may in fact, increase losses of another form, resulting in little net reduction in P loss or water quality improvements at the watershed level.

The more complex agricultural P management strategies have become, the more challenging it became to evaluate the cost-efficiency of the BMPs' and GAPs' implementation. The construction and establishment of agricultural P management practices (both BMP and GAPs) and determining their cost-effectiveness can take several years. Moreover, replicate studies are often difficult to establish due to financial constraints, site-specific characteristics, and temporal variability. To address these complex issues, Veith et al. (2003) developed 'optimization as performance-based method' for AMPs selection and implementation. He argued that this method enables consideration of spatial variation across multiple variables and the assessment of scenario effectiveness for different AMPs and their interactions and as such has many advantages over targeting method (CSA-based approach).

A number of other mathematical models have been developed with the aim to elucidate costs and/or cost-effectiveness of AMPs implementation. One of the most popular and widely used is 'a Soil and Water Assessment Tool (SWAT)'. Since it was developed in the 1990s as a river basin scale mathematical model to quantify the impact and cost-effectiveness of AMPs' implementation in large complex watersheds, it has been described in over 1000 peer-reviewed articles. Over time it had been recognized that SWAT has many limitations however, it continues to be promoted and used as the AMPs cost-effectiveness assessment tool worldwide (SWAT 2018).

### 3.5.9   Challenges in Assessing Treatment Efficiency of AMPs

Despite considerable investments in AMPs' implementation over the past four decades, to date, there has been very little reduction in agricultural P pollution and/or improvement in agricultural water quality, in particular on a watershed scale (Sharpley et al. 2009, 2015; Kleinman et al. 2011; Jarvie et al. 2013; Withers et al. 2014b, 2015).

The greatest challenge in determining treatment efficiency of agricultural P AMPs lies in the extreme complexity of the very problem they are supposed to solve, e.g. pollution originating from many diffuse sources. Diffuse P pollution is a combination of livestock and cropping systems, agricultural surface and subsurface runoff, their field level interactions (both temporal and spatial), and climate (storm frequency and hydrology, temperature). In addition, there are contributions from other rural sources, such as runoff from septic systems leachate fields, sewage treatment facilities, urban development, forestry and industry, and legacy P (Sharpley et al. 2009). Therefore, controlling this type of contamination requires integration of scientific, technological, socio-economical, and educational factors (Novotny and D'Arcy 1999). The evaluation of AMPs' treatment efficiency is further hindered by the issues of scale and the fact that they are implemented on individual farm scales whilst water quality improvement is assessed at the watershed scale (Jarvie et al. 2013; Kleinman et al. 2015; Sharpley et al. 2015).

Jarvie et al. (2013) examined major reasons for the lack of direct evidence of watershed-scale water quality improvements following the implementation of point and NPS AMPs. These included: (i) inadequate intensity and targeting of source and transport BMPs; (ii) inadequate monitoring before and after AMPs are implemented; (iii) complex and lagged ecological responses arising from multiple (physical, chemical, and biological) stressors; (iv) a range of 'complicating factors' with increasing scale from the field to the watershed, including the confounding effects of multiple and complex P sources; (v) biogeochemical buffering and hydrological damping; (vi) wide variations in water and sediment residence times that connect nutrient sources in to the watershed outlet. In addition, the resultant storage, gradual accumulation, and subsequent release from chronic sources, e.g. 'P legacy' between the field and the watershed outlet (in downslope areas, riparian zones, and wetlands, and in stream and lake sediments), is likely to diminish rapid improvement in water quality after the AMP implementation (Kleinman et al. 2011; Jarvie et al. 2013; Withers et al. 2014b). Moreover, in some watershed systems P accumulation reached such high levels that even if P was no longer added to agricultural systems, there would be a considerable time-lag (years or decades) before any improvements in water quality, or aquatic habitats might become apparent (Jarvie et al. 2013).

In addition to this complexity, traditional government funding had been focused only on AMPs' implementation often without any resources for the evaluation of their field performances. As with any other environmental technology, field assessment of agricultural AMPs requires the purchase, installation, and operation of the monitoring equipment, samples collection, and analyses. The automatic flow sampling equipment is very costly. In the USA for example, the most frequently used are those manufactured by Teledyne ISCO (www.teledyneisco.com) at about US$ ~6000 each. The evaluation of the simplest AMP's performance in pollutant mass reductions (passive filter or a trench with defined inlet and outlet) requires the installation of minimum two pieces (inflow and outflow). In addition, this type of equipment is complex and thus requires the engagement of experienced technicians in order to operate them and collect a large number of water samples for analysis. For this reason, unless funded through special programmes to scientific researchers, AMPs performances are evaluated based on grab sampling only (following storm events). Whilst grab sampling is still valuable for the field data collection in terms of pollutant concentrations, it does not provide any information on the actual pollutant mass loading or achievable reductions by the AMP.

### 3.5.10 AMPs Treatment Efficiency in P Reduction

The SERA 17 website provides fact sheets for 33 Phosphorus BMPs, ranging from Application Timing and Methods and Dietary Phosphorus Levels for Dairy Cows to Filter Strips, Riparian zones, and Conservation tillage. Two-Page fact sheets include definition, purpose, brief explanation of how the practice works, application, limitations, treatment effectiveness, costs, and operation and maintenance for each BMP (SERA 17 2018). However, the information provided for both treatment efficiency and cost of establishment is often vague and incomplete. For example, for constructed treatment wetlands treatment efficiency, the fact sheet states P reductions of 20–90% being on average 42%. They further state that the median cost of establishment ranges from $47 000/ha to $378 000/ha for surface and subsurface flow wetlands, respectively. The Filter Strips fact sheet states that their purpose is to reduce sedimentation of surface water bodies and some runoff pollutants. They also state that 'properly placed' buffers can achieve up to 50% P reduction, however, they do not provide further information on what properly placed buffers are. There is no information about the cost of establishment. Instead, there is a general statement explaining that the cost of establishing a filter strip will vary according to the equipment, labour costs, grading, seed and fertilizer selected (SERA 17 2018). Finally there is no indication on the website which BMPs are recommended for point and which for nonpoint/diffuse pollution treatment. Given the fairly large body of literature that describes BMPs and GAPs (Table 3.5. and previous sections) in the following sub sections only a few, most commonly used practices for P pollution reduction from point and NPS will be reviewed. In addition, a few novel methods that have not reached wider commercial application will be presented.

### 3.5.11 Vegetative Buffer Strips (VBS)

VBS, also known as filter strips, buffer strips, buffer zones, have been long accepted as the most common mitigation measure for nutrient loading prevention from NPPS across the globe (e.g. Dosskey 2002; Polyakov et al. 2005; Heinen et al. 2012; Richardson et al. 2012; Georgakakos et al. 2018; Habibiandehkordi et al. 2018). Nonetheless, reported data show that VBS effectiveness is highly variable, ranging from below zero up to almost 100%, depending on the number of factors such as: VBS width, vegetation, nutrient considered, input load, local hydrogeological conditions, and the time period after installation (e.g. Heinen et al. 2012; Richardson et al. 2012; Georgakakos et al. 2018). Richardson et al. (2012) reviewed the history and performance of the fixed-width buffers and concluded that despite billions of dollars in investment and 30 years of promotion and implementation on the agricultural lands worldwide there has been very little evidence in their efficiency, in particular in P reduction. In addition, given that P adsorption by soils is a finite process, the ability of the VBS to reduce P pollution will diminish with time (Georgakakos et al. 2018).

Dosskey (2002) made a survey of four farms in Nebraska and showed that only 9–18% of the farm buffer area was in contact with a runoff. Similarly, Tomer et al. (2003) analysed 20 000 ha watershed in Iowa and found that 23% of riparian zone cells did not receive any runoff during rainfall, 57% had a contributing area of less than 0.4 ha, and 6% received runoff from more than 10 ha. Polyakov et al. (2005) suggested that the root of the general misconception on the VBS potential efficiency may be associated with the general assumption that riparian areas receive runoff in a uniform sheet flow under which they would exhibit maximum buffer efficiency. Sharpley et al. (2009) reported variability in VBS effectiveness in P reduction to range from 40% to 93%, whilst Rao et al. (2009) estimated that it can range from −59% to 56%. In April 2010 a scientific workshop was held in Ballater, UK which brought together 45 scientists, catchment practitioners, and regulators from 15 EU countries to discuss future directions of riparian buffers and the ways to improve their implementation and management. The summary of these discussions and research findings were published in a special edition of the *Journal of Environmental Quality*, volume 41, 2012 (e.g. Heinen et al. 2012; Roberts et al. 2012; Stutter et al. 2012). Yet despite uncertainty and the lack of scientific evidence on the VBS effectiveness in P reduction, they continue to be promoted as the major solution to water quality impairment caused by agricultural pollution. For example following a 2015 report by the Minnesota Pollution Control Agency (MPCA) stating that only three of 93 streams in agriculture-dominant southwest Minnesota fully support aquatic life and only one is considered safe for swimming, their major proposal to address the problem was to create more and wider buffer strips along the state's waterways (Bugbee 2015). A similar approach had been taken in Quebec, Canada, and Vermont to address algae blooms that occur throughout every summer in the Lake Champlain Missisquoi Bay. In Quebec, Lauzier (Ministry of Agriculture, Fisheries and Food) and co-workers from the Institute of Agricultural Research and Development (IRDA) were awarded a 2 million $CAD Funding for 'Projet La lisière verte' to combat diffuse P pollution in the Missisquoi Bay watershed (2007–2009). They proposed installation of croppable, systematic buffer strips with perennial plants, extending the width from a traditional 3 to 9 m (10 to 30 ft) under the slogan 'farm the best, buffer the rest' (Michaud et al. 2009).

Farmers received $CAD 675 /ha for the implementation of the buffer strips during two years of the project. However, the implementation of the buffer strips and prolongation of their width achieved <18% in total P reduction (Hebert 2013).

### 3.5.12   Constructed Wetlands (CW)

CW have been used extensively and around the world for both point (agricultural effluents) and nonpoint (surface and subsurface runoff) mitigation. However, it has been universally accepted amongst the CW scientists and engineers that their performance in removing P from these sources has been poor regardless from the complexity of design used, especially in cold climates (e.g. Hunt and Poach 2001; Luderitz and Gerlach 2002; Weber et al. 2007; Drizo et al. 2008a; Vymazal 2010). Knight et al. (2000) compiled the Livestock Wastewater Treatment Database for North America containing treatment performance of 38 CW systems. They reported that average TP performance averaged for all livestock management CWs (including cattle feeding, dairy, poultry and swine) was only 42%. Drizo et al. (2008a) investigated performance of four subsurface flow CW in treating combined milk parlour and barnyard effluent from ~120 cows at the University of Vermont dairy farm, USA. They found that a non-aerated CW dissolved P removal efficiency was only 19.9% during a three years period of research (2004–2007). Furthermore, although supplemental aeration increased P removal efficiency, it remained below 50% during the entire period of investigation (Drizo et al. 2008a). In a similar project, Forbes et al. (2011) investigated P removal efficiency of a CW designed to treat combined milking facilities and manure runoff from a 170-cows farm located at College of Agriculture, Food and Rural Environment in Northern Ireland. They reported unusually high CW P removal efficiency, averaging 95% over five years of investigation. However, this particular CW design consisted of five large ponds with extremely long residence times (100 days) and the entire system covering $12\,510\,m^2$ (Forbes et al. 2011) as opposed to system in VT which covered $880\,m^2$ (Drizo et al. 2008a). Kadlec (2016) reviewed large CW for P control which included 66 systems with a median size of 210 ha ($2\,100\,000\,m^2$). He pointed out that although these very large CW achieved 71% P reduction in average, thanks to at a low median hydraulic loading of only 2.55 cm/day, the amount of P stored (phosphorus load removed, PLR) has been low, with a median of just $0.77\,g\,P/m^2/year$. Moreover, CW reviewed were of different age which should be taken into account. Phosphorus removal will often be higher in the first few years of operation and then decrease gradually over time as the soil adsorption sites become saturated (e.g. Richardson and Craft 1993; Drizo et al. 2008a; Rozema et al. 2016).

In the past several years there has been increasing interest in the use of CW for agricultural tile drainage treatment and pollution reduction. This research has been reported in 10 research papers in a special issue of *Water*

(ISSN 2073-4441), published online in June 2018 (*Water* 2018). Based on the long-term research on CW and agricultural tile drainage, the New Zealand National Institute of Water and Atmospheric Research (NIWA) developed Guidelines for Constructed Wetland Treatment of Tile Drainage. The purpose of the document is to guide farmers, farm advisors, rural contractors, and regional council staff to appropriately locate, size, design, and construct effective treatment wetlands (Tanner et al. 2010).

### 3.5.13  Phosphorus Removal from Agricultural Tile Drainage

McDowell et al. (2008) and Drizo (2010) investigated the use of industrial by-products to mitigate P pollution from tile-drained land. In New Zealand, McDowell et al. (2008) suggested that backfilling tile drains with IMS and a small proportion of basic slag could be effective means of decreasing P loss from high P soils. In the USA, Drizo and co-researchers investigated passive filter systems filled with EAF SSA for treatment for P reduction from agricultural tile drainage in several pilot projects in Vermont and OH (Drizo 2012). Buda et al. (2012) reviewed emerging technologies low-cost P sorbing materials for removing NPS P from surface water and groundwater. They pointed out that high flows occurring in tile drained land can overwhelm the saturated hydraulic conductivity of the filter material, resulting in decreased retention times and consequently reduced treatment efficiencies. Furthermore, they suggested that the application of filtration technologies for concentrated flows may be limited to treating lower flow rates or that flows control structures may need to be installed prior to a filter system Buda et al. 2012). Drizo developed the very first interim conservation practice standard for P removal from surface and subsurface flows, known as the Phosphorus Removal System #782 (Chapter 2, Case study 3). The Standard recommends that the media should have a PRC of at least 0.50% by weight of materials, or 4.5 kg P/ton of media. It also underlines that the particle diameter of the media ought to provide sufficient permeability for the anticipated flow (e.g. USDA NRCS Vermont 2013). In 2015, the Friends of Northern Lake Champlain (FNLC) and Stone Environmental collaborated on the first implementation of P removal system #782 in Vermont. They compared performance of the two locally available media: (i) alum-based drinking water treatment residuals (DWTR) and (ii) crushed limestone from Swanton Lime in Swanton, Vermont (Brown 2017). Contrary to the laboratory testing that indicated high PRC by the two selected materials, P reduction performance in the field was very poor. In addition, there has been progressive clogging of the media. Brown (2017) attributed poor system performance to the poor selection of the system location (absence of elevation) and associated engineering and high-flow events. Whilst both the absence of elevation and high-flow events can cause under performance, the selection of a right media, in particular appropriate diameter size, is of the utmost importance. Brown (2017) did not provide any details on the media particle sizes or P retention capacities determined in the laboratories.

During late 1990s, overestimates of P retention capacities measured in laboratories led to a failure of a number of full-scale studies that investigated the use of CW for P removal (Drizo et al. 2002).

In September 2017, the very first P removal system #782 based on basic furnace slag (BFS) was installed in Sheboygan County, WI (Figure 3.6). Whilst the system initially showed 66% reduction in dissolved and total P, respectively (autumn 2017), the efficiency deteriorated drastically due to similar reasons experienced in VT, e.g. absence of elevation. Following adjustments at the site the efficiency of the system improved to 45% during October 2018 (E. Fehlhaber, personal communication).

Chad Pen and co-workers took the well-established research on passive filters for P removal based on industrial by-products and branded the commonly used term 'P removing filters' into '*P removal structures*' (e.g. Penn et al. 2012, 2017; Penn and Bowen 2018). Similarly, they branded the accustomed term '*industrial by-products for P removal*' into P sorption materials '*(PSMs)*'. Finally they took an over a decade old and recognized method for presenting filters' efficiency in P removal (plotting cumulative P added versus cumulative P adsorbed, e.g. Drizo et al. 2002, 2006; Weber et al. 2007) into '*P removal curve*'. In their recent review of Phosphorus Removal Structures they attempted to provide an assessment for comparison of 'P removal structures' using data from a variety of CW and filters for P removal. In the conclusions of the paper, they stated that the performance of P removal structures should be presented 'as cumulative P removal (mass of P per mass of media) as a function of cumulative P loading' (Penn et al. 2017). *It should be noted that their statement was published 15 years after this type of data presentation has been used by other researchers* (e.g. Drizo et al. 2002, 2006, 2008b; Weber et al. 2007).

According to the Cost869 Project (Box 3.3.), in Europe, practices recommended for nutrient loss (e.g. pollution reduction) from agricultural water

**Figure 3.6** Phosphorus Removal System #782 in Sheboygan County, WI showing installation and completed construction along with Agridrain water level control structures.

management involve measures to control drainage and irrigation (http://www.cost869.alterra.nl). For drainage control, practices are categorized on those focused on reduction of (i) overflow to surface water (e.g. construction of ponding systems and grassed waterways, installation of sedimentation boxes); (ii) subsurface flow to surface water (removal of trenches and ditches to allow field drainage systems to deteriorate and installation of artificial drains to improve sub-surface drainage systems); and (iii) loss by artificial drainage to surface water (e.g. implementation of controlled drainage for reducing the amount of water leaving a field; letting tile drainage water to irrigate meadows/interrupt artificial drainage). For nutrient pollution control from irrigation, nutrient loss with surface irrigation and water recovery on irrigated fields for water and nutrient cycling are listed as good practices. For each of the listed practices two-page fact sheets are provided on the website. Similar to SERA17 Fact sheets, they include short description, rationale, applicability, effectiveness, time-frame, and costs. In addition Cost869 fact sheets also include potential environmental side-effects, but do not address costs of operation and maintenance.

The latest report on the BEMP for the agriculture sector (European Commission 2018a) lists the following measures as BEMP to mitigate tile drainage pollution impacts: contour ploughing, break slopes, cultivation of tramlines, avoidance of compaction, low ground-pressure impact tyres on vehicles, and erosion risk planning. However, whilst all of these recommended measures may improve soil infiltration and therefore aid in reducing surface flows, their contribution to minimizing P loading has not been quantified (European Commission 2018a).

### 3.5.14  Phosphorus Removal Methods from Livestock Production

#### 3.5.14.1  Animal Manure (Solid)

Manure is usually managed as dry solid manure or liquid slurries, stored in especially designed storage facilities or structures. Liquid manure and wastewaters are sent to detention ponds or lagoons for settling out the solids fraction and reducing the volume through evaporation. Lagoons also serve as a temporary storage facility for land application. However, the quantities of manure generated on the confined animal operations often exceed local crop needs and areas available for application, posing considerable challenges in P management (e.g. Sharpley et al. 1994; Sims et al. 2005; Sharpley et al. 2006; Teenstra et al. 2014). This is particularly the case in the USA, Canada, and other temperate regions of the world. Vermont, USA introduced a winter manure spreading ban for a period of six months (15 December to 1 April). In addition, in many areas manure has been stored in open pits that can still cause significant P pollution at each precipitation event (Sharpley et al. 2006; Teenstra et al. 2014).

A number of AMPs were developed with the aim to reduce and control the quantity of P in manure showing varying efficiency. For example, feed P management via manipulation of dietary P intake by animals (e.g. Rotz et al. 2002; Dou et al. 2003; Sharply et al. 2006; Wang et al. 2014) revealed P reductions in manure ranging from 5% to 60%, depending on the diet method used (Sharply et al. 2006).

Chemical amendments have been investigated for their potential to reduce P solubility in dairy and swine manures. Anderson et al. (1995) found that gypsum application to dairy manure-amended soils reduced soluble P between 40% and 60%. Dou et al. (2003) tested combustion fly ash and alum and found that they reduced P solubility of manure by 80–99% and 50–60%, respectively. In their review of products used as manure chemical amendments for reducing P losses to environment, Torbet et al. (2005) stated that gypsum ($CaSO_4$), ferrous sulphate ($FeSO_4$), and lime ($CaCO_3$) were particularly promising, in particular as all three of them are considered waste products. However, despite the number of approaches and methods tested over the years, to date none have been shown to be economically viable for farmers to be adopted as a P management practice (Sharply 2015).

Physical treatment based on traditional municipal wastewater effluents treatment methods using coagulants and flocculants were also tested. However, use of these methods does not alter the total amount of phosphorus that needs to be handled (Sharpley et al. 2006).

More recent options to handle solid manure include anaerobic digestion, and composting and liquid–solids separation. However none of these technologies remove P from the pollution source. Anaerobic digestion concentrates P in solid or liquid products. Thus this technology will be described in Chapter 4 on Phosphorus Recovery. Composting can become an additional source of P through runoff and leachate. Practices recommended by SERA17 for the USA (SERA 17 2018), Cost869 Project (Box 3.1.) and the Joint Research Centre of the EC (European Commission 2018a), which target P reduction specifically are listed in Table 3.7. However, general practices such as for example: site solid manure heaps away from watercourses and field drains; site solid manure heaps on concrete with collection of the effluent; manure production and application management (e.g. increase the capacity of farm manure [slurry] stores, adopt batch storage of solid manure or slurry, minimize the volume of dirty water produced, compost solid manure, change from slurry to a solid manure handling system, incorporate manure into the soil); and manure surplus management are excluded. This is because although they can contribute to reduction in P loading from animal manure, they are general management practices and were not specifically developed to mitigate P pollution from this source.

### 3.5.14.2 Livestock Liquid Effluents and Runoff

The fact sheets describing practices for Livestock Liquid Effluents and Runoff management are listed at SERA17, Cost869, and Living Water Exchange (2018) websites. They include general practices such as Farmyard/

Table 3.7 Agricultural management practices for P removal from animal manure.

| Practice Fact Sheets | P Reduction Efficiency (%) | Source |
| --- | --- | --- |
| *Manipulation of dietary P uptake* | | |
| Dietary Phytase to Reduce P Losses from Animal Manure | 15–25% compared to manure from animals provided conventional diets. | SERA 17 (2018) |
| Dietary P Levels for Dairy Cows | Not provided. Limited descriptive information of potential P reduction in a runoff based on experimental trials | SERA 17 (2018) |
| Composting Effects on P Availability in Animal Manures | Trials with rain simulator showed that P losses from compost application were at least 50% less than from manure. | SERA 17 (2018) |
| Reducing of N and P in dairy Nutrition | Not provided | Cost869 (2018) |
| The Impact of Nutrition on Reduction of P Excretion in Pigs | Results from one study indicated that the use of high energy diets could reduce P-excretion by 63%. | Cost869 (2018) |
| *Chemical amendments* | | |
| Treating Poultry Litter with Aluminium Sulphate (Alum) | 75–85% from small watersheds and small plots, respectively | SERA 17 (2018) |
| Treating Swine Manure with Aluminium Chloride | Not provided | SERA 17 (2018) |
| Phosphorus Immobilizing Amendments to Soil | Not provided | Cost869 (2018) |
| *Liquid–solids separation* | | |
| Physical Manure Treatment (Solids Separation) | The effectiveness in removing nutrients in the solid fraction is not as well documented. Limited data showed up to 35% reduction in P. | SERA 17 (2018) |

Feedlot Runoff Management (e.g. various systems for collection, treatment. and reduction of runoff), 'Milkhouse Filters', which are defined as vegetated and non-vegetated infiltration strips and septic type systems (SERA 17 2018) and 'minimize the volume of dirty water produced' (Cost869 2018). However, none of these have been developed to target P reduction.

Between 2004 and 2012, Drizo and co-researchers investigated several sustainable practices for livestock farm effluents and runoff management which specifically targeted P reduction in northern, cold climates (northeastern USA). These included CW (e.g. Weber et al. 2007; Drizo et al. 2008a; Tunçsiper et al. 2015), P removing filters with SSA as the media (e.g. Weber et al. 2007; Drizo et al. 2008b, Bird and Drizo 2009) and integrated CWs and P removing filters (Lee et al. 2010; Adera et al. 2018). This long-term research revealed that full-scale CWs retained less than 20% of the P during the first three years of operation, further confirming this technology's ineffectiveness in P removal in cold climate countries (Drizo et al. 2008a). More importantly, this pioneering research provided substantial scientific evidence on the superior performance of SSA based filters (Lee et al. 2010; Adera et al. 2018). In addition, between 2009 and 2011, Drizo and co-researchers also conducted a comparative study, the first of its kind, which provided the first field-based technical and economic assessment of the novel SSA P filters technology and its contribution to the treatment efficiency of the two (grass filter area treatment, GFA and vegetative filter strip, VFS) other conservation practices currently used in VT and other States (Drizo 2011). Water quality analysis revealed that current accepted conservation practices are costly and inefficient from silage leachate runoff and that the addition of a single SSA berm could considerably improve dissolved P, pathogens, and organic matter reductions. The economic cost-benefits analysis showed that the cost of a Vermont NRCS Standard for Vegetated Treatment Area (Acre) Code 635 which is defined as 'an area of permanent vegetation used for agricultural wastewater treatment' is $13 170.15 for a bunk silo 1 acre in size, for the estimated life span of up to 15 years. Drizo and co-researchers made recommendations to the USDA that the implementation of an innovative 'treatment train' system consisting of a SSA P removing filter, a single vegetative treatment area and a trench filled with P adsorbing material and vegetated with local grasses, would be considerably cheaper (saving in average $3000/acre to the farmer) and have significantly higher performance in pollutant reduction (Drizo 2011).

### 3.5.14.3 P Reduction Efficiencies of Other Agricultural BMPs

Sharpley et al. (2009) summarized P reduction efficiencies in surface runoff for 15 different BMPs reported in the literature between 1992 and 2005 including source measures (P rates balanced to crop use, subsurface applied P versus surface broadcast and adoption of nutrient management plans) and transport measures (no till versus conventional tillage, cover crops, diverse cropping systems and rotations, contour ploughing and terracing, conversion to perennial crops, livestock exclusion from streams versus constant intensive grazing, in-field vegetative buffers, sedimentation basins, riparian buffers, and wetlands). They pointed out that conservation practices vary substantially in effectiveness within and amongst watersheds with P reduction efficiencies ranging from 7% to 63%, contour ploughing 30% to 75%, livestock exclusion 32% to 76%, riparian vegetative buffers 40% to 93%. The greatest variability was found in wetlands (0–79%), managed grazing versus constant intensive grazing (0–78%) and subsurface applied P versus surface broadcast (8–92%). They also underlined the need for robust water quality monitoring and concluded that whilst there are some effective BMPs, none of them should be seen or used individually as the primary mechanism by which a farmer would reduce P losses.

Rao et al. (2009) used a Variable Source Loading Function (VSLF) model, which captures the spatial and temporal evolutions of variable source areas (VSAs) in the landscape to evaluate the effectiveness of eight different BMPs implemented on a 164-ha farm in the New York City source watersheds in the Catskills Mountains of New York State, and monitored over 11 years. These included VBS, nutrient management plans, strip cropping, crop rotation, filter strips, culvert crossing, animal waste management, and barnyard management. Similarly to Sharpley et al. (2009) all of the practices showed large variability in effectiveness, with the highest found in animal waste management (−117% to 40%), nutrient management planning (−66% to 94%) and filter strips (−56% to 59%). A large number of scientists have used mathematical modelling and optimization research tools in attempts to evaluate agricultural BMPs' effectiveness in a watershed context (e.g. Randhir et al. 2000; Srivastava et al. 2002; Hsieh and Yang 2007; Gaddis et al. 2014; Xie et al. 2015).

### 3.5.14.4   Recent Research Changes

Recognizing the cost and uncertainty in the BMPs' performances in P reduction, over the past several years there has been a positive shift in AMPs research from the mere BMPs implementation towards the investigations of the factors that affect BMPs performances and adoption leaders in the field (e.g. Kleinman et al. 2011; Sharpley et al. 2011; Jarvie et al. 2013; Withers et al. 2014b) started to bring attention to the chronic sources of P accumulated in soils over the years, and named 'legacy phosphorus'. For example, Kleinman et al. (2011) stressed the importance of distinguishing between acute, temporary (recently applied P as manure/fertilizer) and *'legacy phosphorus'*. They also proposed adjustment of the agricultural management to address the role of hydrology on P transfers and to other conservation practices (e.g. tillage management); taking into account subsurface transport of P (tile drainage and tillage management) and P legacy (recognizing the need for novel practices to confront P legacy); the need to control P in pasture-based systems. In addition, they recognized the potential of novel practices to intercept surface and subsurface farm flows (drainage lines and ditches) and adjacent to areas of acute P accumulation potential (e.g. barnyards, subsoils). Finally, they underlined that despite recommendations that the CSA should be targeted for cost-effective remedial management, the actual identification of the places where P mobilization is likely to occur within CSA is a difficult task. Sharpley et al. (2011) described a paradigm shift in agricultural P management strategies from unilateral recommendation of conservation measures to address P loss across a watershed towards specific targeting (both spatial and temporal) of particular management practices on CSA within a watershed. They also pointed out that CSA identification cannot be regarded as the only tool in mitigating P pollution from agricultural sources and that other long-term factors such are farm and watershed scale P imbalances, legacy P sources, and vertical stratification of P in soils must be taken into consideration. In addition, they reviewed Phosphorus Index, a site assessment tool which was designed and proposed 20 years ago with the aim of identifying and ranking CSA of P loss based on site-specific source factors (soil P, rate, method, timing, and type of P applied) and transport factors (runoff, erosion, and proximity to streams) and has been adopted in as many as 47 US States. They pointed out that in some cases BMPs implemented to reduce one form of P loss may in fact, increase losses of another form, resulting in little net reduction in P loss or water quality improvements at the watershed level (Sharpley et al. 2011).

In September 2013, 150 leading experts in the field of agricultural P management attended a workshop in Uppsala, Sweden and discussed future research needs and directions. Major findings and strategies for improvement in agricultural P management were published in a special issue of *Ambio* journal (volume 44, Issue 2 Supplement, March 2015) 'Future agriculture with minimized phosphorus losses to waters'. Sharpley et al. (2015) summarized the key topics of discussions and for each of them proposed research needs and directions. These included: (i) P management in the changing world; (ii) transport pathways of P from soil to water; (iii) monitoring, modelling, and communication; (iv) importance of manure and agricultural production systems for P management; (v) identification of appropriate mitigation measures for reduction of P loss; and (vi) implementation of mitigation strategies to reduce P loss.

However, none of these papers discussed issues related to the obstacles to the innovation and introduction of the novel conservation practices to the agricultural market, and/or considered the factors that promote or impede retrofitting of the already implemented, but under performing BMPs and GAPs.

### 3.5.15 Obstacles to Innovation

To better understand the factors that severely impede innovation in agricultural P pollution mitigation one should consider several crucial facts: (i) since their inception in 1950s, the BMPs, in particular VBS have been promoted and implemented across the world despite a substantial lack of evidence on their actual effectiveness; in the USA for example, (ii) funding for agricultural water quality protection has been focused solely on BMPs implementation, without any financial assistance for the water quality monitoring which is a prerequisite for the evaluation of their effectiveness; (iii) severe limitations in research funding (e.g. funding is usually offered only for pilot/demonstration projects whilst for the inclusion as an 'interim conservation practice' the regulators require data on the full scale performances; for most of the research grants calls the project duration is limited to 12 months whilst regulators require to see 'long-term' data; water quality monitoring costs are not eligible in most cases; in the programmes that allowed novel technology investigation, 1:1 funding match is required which is very difficult to find as the investors are interested in fast, short-term return on investment [ROI] and are generally not interested to invest in solving agricultural P pollution problems); (iv) the complexity and length of approval process by the regulators (e.g. in the USA, NRCS and agencies of agriculture staff) for potential inclusion of novel conservation practices; (v) the absence of certification standards in particular for P removal practices/technologies (demanding more time, more research, more data, and extending the process to 3–5 years); (vi) farmers social behaviour, e.g. even in a rare instances when farmers are genuinely interested in implementing novel solutions for P reduction, they fear that if they provide their land for new research/practice and a solution is found, the regulators would impose obligatory implementation on all farms; (vii) general absence and inadequacy of agricultural phosphorus pollution legislation and luck of enforcement.

## 3.6 Phosphorus Removal from Urban Stormwater Runoff

### 3.6.1 Urban Stormwater Runoff Treatment – Background

The combined sewer collection systems, using single-pipe systems that collect both sewage and urban runoff from streets and roofs were introduced in the late nineteenth century. As the majority of the cities at that time did not have sewage treatment plants, there was no perceived public health advantage in constructing a separate storm sewer system (US EPA 2004; Tibbetts 2005). The cities' rationale was that it would be cheaper to build a single, combined sewer system (CSS) instead of using urban cesspool ditches. However, when too much stormwater is added to the flow of raw sewage, the result is frequently an overflow. Therefore, over the past several decades, the combined sewer overflows (CSOs) have become the focus of a debate regarding the best techniques to manage growing volumes of sewage and stormwater runoff in urban communities across the globe.

Concomitantly, and similarly to agricultural BMPs, a number of the 'stormwater BMPs' have emerged, using various terminology to describe very similar techniques proposed to mitigate stormwater runoff flows (Box 3.4). The most common BMP's used are detention/retention ponds, exfiltration ponds, sand filters, recirculating filters and various bio-retentions areas (grass waterways, rain gardens). However, none of these stormwater prac-

---

Box 3.4 Terminology Surrounding Urban Drainage

Fletcher et al. (2015) provided a comprehensive description of the evolution and application of terminology surrounding urban drainage. They provided a brief history and overview of each of the terms used: (i) the low impact development (LID), the most commonly used in North America and New Zealand; (ii) water sensitive urban design (WSUD), term used in Australia; (iii) Sustainable urban drainage systems (SUDS) or sustainable drainage systems (SuDS), terms most commonly used in the UK; (iv) Alternative techniques (ATs) or compensatory techniques (CTs), terms used in French speaking countries; and (v) BMPs, terms used in the USA and Canada. In addition, they also described the term green infrastructure (GI), which emerged in the USA in the 1990s and whose concept goes far beyond stormwater, e.g. it includes landscape architecture, landscape ecology and the provision of ecosystems services.

tices were developed with the aim to reduce nutrient pollution. Their major purpose is to provide storage for excessive storm water volumes occasionally, also treatment of organic matter, suspended solids (SS), and bacteria.

### 3.6.2   The International Stormwater BMPs Database (ISBD)

In early 2000, the American Society of Civil Engineers (ASCE) and the USEPA initiated a comprehensive project to create an international database of urban stormwater runoff BMPs. In 2004, the project transitioned to a coalition of partners which are today led by the Water Research Foundation (WRF), including the Federal Highway Administration (FHWA), American Public Works Association (APWA), and the Environmental and Water Resources Institute (EWRI) of ASCE (BMP database 2018). Today, the ISBD project website features over 700 BMP studies, performance analysis results, tools for use in BMP performance studies, monitoring guidance, and other study-related publications (BMP database 2018).

However, despite being very comprehensive, the ISBD has a number of shortcomings. The biggest drawback is in the fact that data include only BMPs' influent and effluent concentrations and there is no information regarding the influent and effluent loads and potential pollutants mass reductions achievable by these BMPs (BMP database 2018). Another shortcoming is that the website is difficult to navigate, e.g. it does not provide a search engine that would enable the user to find information using key words. Whilst the main 2018 BMP database document containing BMPs performances contains a box for the key words search, the attempt to find information on terms nutrients, phosphorus and/or performance do not provide any results (BMP database 2018).

The 2016 Summary Statistics, which can be found under urban stormwater research reports tab, provide data on influent and effluent total P concentrations for 11 different BMPs, classified by the categories. The BMP categories included are grass strips, bioretention, bioswales, composite/treatment train BMPs, detention basins (surface/grass-lined), media filters (mostly sand filters), porous pavement, retention ponds (surface pond with a permanent pool), wetland basins (basins with open water surface), a combined category including both retention ponds and wetland basins, and wetland channels (swales and channels with wetland vegetation). According to the median inlet and outlet values shown, P removal efficiency of these BMPs were generally low. The highest removal efficiency achieved by the retention pond (55%) followed by the wetland basin/retention pond (44%), porous pavement (42%) and media filter (40%). Bioretention exhibited the worst performance (−84%), followed by grass swale (−67%) and grass strip (−21%). This is in accordance with a number of researchers who reported highly variable, and generally low performance of bioretention systems in P removal (e.g. Dietz and Clausen 2005; Hunt et al. 2006; Roy-Poirier et al. 2010; DeBusk and Wynn 2011). The underperformance of the stormwater

BMPs in P reduction could be attributed to the fact that none of these practices were originally developed with the aim to provide this function.

During the past several years there has been an increasing number of studies that investigated ways to improve the performance of stormwater BMPs using a variety of 'Performance Enhancing Devices' (PEDs) including, but not limited to the 'engineering media' (natural and industrial by products, described earlier) incorporation in bioretention systems, rain gardens, CW, sand filters, infiltration trenches, and retention/detention ponds retrofitting with passive P filters (e.g. Davis et al. 2009; Ma et al. 2011; Ahiablame et al. 2012; Drizo 2012; Gulliver 2014; Hirschman et al. 2017).

Media tested included Iron- or Aluminium-based materials (iron filings, steel wool, SSA, water treatment residuals (WTRs), acid mine drainage (AMD) residuals, fly ash) and carbon-based materials such as biochar, wood chips, and activated carbon. However, Hirschman et al. (2017) underlined that the use of compost and other organic matter in bioretention media can result in nutrient leaching and should be avoided in higher concentrations within the bioretention media.

### 3.6.3   Commercial Products

Given the steady increase in eutrophication and P pollution problems a number of new products have emerged on the stormwater market in the past 10–15 years claiming P and sediment removal efficiencies of nearly 90% 'using a variety of sustainable media'. Some of the most popular products include high technology, expensive, cartridge based products such as: (i) Contech Engineered Solutions Stormwater Management StormFilter®, based on PhosphoSorb™, a lightweight media built from a Perlite base (Contech Engineered Solutions 2018), (ii) Imbrium Stormwater Solutions, based on 'SorbtiveMEDIA' (Imbrium Systems 2007, 2010), and (iii) Hydro International 'Up-Flo Filter' based on Hydro International CPZ mix™ media (consisting of activated carbon, pumice, and zeolite aggregates mix) (Hydro International 2009, 2018). However, these are cartridge based products housed in underground vaults, are highly complex and costly, and require high maintenance and are prone to clogging (Restoration and Recovery 2014).

In addition to these high-tech complex solutions, a few other biofiltration systems emerged on the market, directly aligned with the LID design guidelines. These novel biofiltration systems consist of a tree or a shrub, planted in the organic media (composed mainly of the soil, peat, and wood chip mix), and especially designed to fit the surrounding landscapes. One of these products had been housed in a concrete container, and offered by Filterra–Americast, Inc. company. However, in 2014, Filterra–Americast, Inc. was acquired by Contech Engineered Solutions (Informed Infrastructure 2014) who became a provider of these products (Contech Engineered Solutions 2018). The other product is FocalPoint High Performance Modular Biofiltration System offered by Convergent Water Technologies (CWT), a company which specializes in delivering the 'next generation' LID technologies (Convergent Water Technologies 2018). The key in the FocalPoint's high performance in organic matter and SS removal from storm water runoff is the biofiltration media, which was developed by Larry Coffman, the creator of the LID in the USA (Coffmann et al. 1999; Coffman 2000). In January 2014 CWT obtained an exclusive licence from Water and Soil Solutions International to use their PhosphoReduc recyclable phosphorus filtration media in FocalPoint and other CWT's products that aim to provide P reduction and harvesting (Convergent Water Technologies 2018; Water and Soil Solutions International 2018).

However, given the lack of regulatory requirements for P removal from storm water runoff, and the costs of environmental certification for the novel storm water products, to date, only one Focal Point-PhosphoReduc system was installed as a demonstration project, on a Mills Creek Golf Course in Sandusky, OH. Due to the luck of funding for automatic sample collection and analysis, the system was evaluated only via grab sampling which has not been sufficient in order to place this novel product on the stormwater market.

### 3.6.4  Challenges in Bringing Innovative Products to the Stormwater Runoff Management Market

Similarly to other market segments, the greatest obstacle in bringing innovative products to the market is the cost of environmental technology verification (ETV) programmes (Box 3.5). In a case of P removal products, the lack of regulatory requirements represents an additional, unsurmountable obstacle as it diminishes any potential interest from investors.

In the USA there are two major certification programmes for the evaluation and approval of stormwater BMPs that include protocols for the assessment of P removal efficiency: (i) the Washington State Department of Ecology's Technology Assessment Protocol – Ecology (TAPE) Program, a process for evaluating and approving emerging stormwater treatment BMPs (Washington Stormwater Center 2018); and (ii) the New Jersey Corporation

---

Box 3.5  The Cost of Stormwater Products Testing, USA

According to the National STEPP Workgroup Steering Committee, field studies are estimated to cost between $250 000 and $700 000 depending on the number of storm events needed to meet the testing protocol requirements. The average cost of TAPE Program is $250 000–$350 000; the costs of the newly developed VTAP programme is in the same range, e.g. $312 000 (24 storm events required, at the cost of about $13 000 per storm event), with the costs reaching $500 000. It had been estimated that it would take the manufacturer five to seven years to recover from the costs of VTAP testing (WEF 2014).

for Advanced Technology, NJCAT (NJCAT 2018). A brief history and description of the USA National Stormwater Testing and Evaluation for Products and Practices (STEPP) Initiative has been provided by the Water Environment Federation (WEF 2017).

There are several protocols that have been developed for particular States and are not transferable. These include CALTRANS (California), Georgia Technology Acceptance Protocol (GTAP), Massachusetts Stormwater Technology Evaluation Project (MASTEP) and Virginia Technology Acceptance Protocol (VTAP) (WEF 2014).

Protocols for obtaining the ETV in Canada and Europe can be found on their respective websites (ETV Canada 2018; European Commission 2015e). However, similar to the USA testing protocols, both Canadian and EU ETV Programmes are extremely expensive and time consuming.

Considering the impacts of climate change on rainfall intensities and stormwater runoff peak flow and volumes, there is an urgent need to develop climate friendly, multifunctional stormwater practices and GI. In order to achieve this goal, it is absolutely critical to modify the existing, and/or develop novel protocols which would be internationally recognized and enable more rapid evaluation and adoption of novel stormwater practices.

## 3.7  In-Lake Treatment of P

In addition to methods and technologies to prevent, reduce, and mitigate external loading of P caused by human activities it is of a vital importance to control the internal loadings that may be caused by release of P from the lake sediments under certain environmental conditions (e.g. Wu et al. 2014; Chapter 1).

Various remediation strategies have been developed to control internal P loading in lakes, which can be broadly divided into in situ and ex situ techniques (e.g. Spears et al. 2014; Zhang et al. 2016; Wang and Jiang 2016). The most commonly used ex situ technique is sediment dredging, despite its high costs ($132–1750/m$^3$) and the fact that during the process, the re-suspended sediments may be released back into lake water (Reddy et al. 2007; Wang and Jiang 2016).

Wang and Jiang (2016) conducted a comprehensive review of the range of physical, biological, and chemical in situ immobilization techniques to reduce the internal P loading from lake sediments for eutrophication control. They discussed a number of chemicals used and classified them into nine categories: (i) aluminium (Al)-based compounds, (ii) iron (Fe)-based compounds, (iii) calcium (Ca)-based compounds, (iv) nitrates, (v) Phoslock (a lanthanum [La]-modified clay), (vi) dewatered DWTRs, (vii) Z2G1 (an aluminized modified zeolite), (viii) chemicals used in combination, and (ix) other chemicals. For each category they described dosages, applications, effectiveness and stability, and advantages and disadvantages. A review of a number of studies lasting from 1 month to 20 years revealed that the Al treatment can remain effective for 8–11 years in polymictic lakes (lakes that have a uniform temperature and density and are too shallow to develop thermal stratification) and for 13–20 years in dimictic lakes (lakes that mix from the surface to bottom twice each year) based on the reduction in total P and chlorophyll a in lake water as well as in the internal P loading rate (Wang and Jiang 2016). A study of ten lakes on Cape Cod, Massachusetts, USA which have been treated with Al for two decades showed that for these lakes, the concentration of applied Al treatment ought to be at least 10 times the mobile P concentration in the upper 10 cm of sediment exposed to anoxia (Wagner et al. 2017). Fe salts have been used to a lesser extent, than Al salts. Also to date there has been no definite recommendations for Fe dosage calculations (Spears et al. 2014; Wang and Jiang 2016). In a case of Ca-based treatment, the research has shown that the addition of $CaCO_3$ (calcite), $Ca(OH)_2$, and gypsum reduced chlorophyll a concentrations either temporarily or long-term depending on the level of P concentrations in lake waters. It had also been found that treatment with $CaCO_3$ (calcite) and lime can control macrophytes due to a rise in pH (Wang and Jiang 2016).

One of the most widespread used products in the recent years has been Phoslock, a lanthanum-modified bentonite which was specifically developed as a P-absorbent by CSIRO Australia (Douglas 2002; Spears et al. 2014; Wang and Jiang 2016). Phoslock had been generally effective in reducing P levels in sediments. However it had been noted that P adsorption by Phoslock was highest at pH 5–7, and that the P adsorption capacity decreases at pH > 9 (Ross et al. 2008).

Although the use of chemical additives is becoming commonly considered as a eutrophication management tool there are still a number of uncertainties regarding the universal acceptance of this method as the eutrophication management tool (Spears et al. 2014; Wang and Jiang 2016). Spears et al. (2014) recently reviewed novel developments in the in lake treatment of P and suggested a number of research needs. These included but are not limited to: (i) development of novel protocols to combine traditional long-term data with high frequency monitoring programmes; (ii) collection of the evidence of the lakes' responses following treatment to provide data for meta-analysis; (iii) an assessment of the environmental effects associated with all (well-established and in particular new) products; (iv) design decision support systems to assist material selection across a range of receiving water types; (v) development of the standard protocols of the likely behaviour of the amendment material in the receiving water and (vi) perform the assessment of the materials' application in the context of 'dose–response' (i.e., repeated smaller doses) as opposed to the common 'single pill' (i.e., single large dose) approach to minimize the risk of unintended consequences.

# 4

# Phosphorus Recovery Technologies

## 4.1 Introduction

A decline of the global phosphate reserves, in particular, those of high grade, has been known for the past 30 years (e.g. Steen 1998; CEEP 1998; Balmer 2004; Roeleveld et al. 2004). However, it was only after the immense price fluctuations on the phosphate rock market in 2007 and 2008, which brought about price increases of more than 900%, that the question of P availability and imminent physical phosphate rock scarcity started to attract increasing attention from both academics and the public (Heckenmüller et al. 2014).

As a result of this renewed concern regarding global P and food security, the researchers at the Institute for Sustainable Futures at the University of Technology, Sydney (UTS), and the Department of Thematic Studies – Water and Environmental Studies at Linköping University, Sweden, established the Global Phosphorus Research Initiative (GPRI) in 2008. Over the past decade, four additional independent research institutes in Europe, Australia, and North America joined the GPRI: the French National Institute for Agricultural Research (INRA), the Stockholm Environment Institute (SEI, Sweden), the University of British Columbia (UBC, Canada), and Plant Research International (PRI) at Wageningen University in the Netherlands (Phosphorus Futures 2018). In parallel, the United Nations Environment Programme (UNEP) launched the Global Partnership on Nutrient Management (GPNM) as a global forum for advancing member science-based nutrient management policies and practices (UNEP 2010). In 2015, the UNEP's GPNM launched its Phosphorus Task Team, and an Action Plan for regional and global stakeholder engagement was identified (Phosphorus Futures 2018).

*Phosphorus Pollution Control - Policies and Strategies*, First Edition. Aleksandra Drizo.
© 2020 John Wiley & Sons Ltd. Published 2020 by John Wiley & Sons Ltd.

A number of research papers were published, warning about depletion of P reserves and the implications on global food security and calling for P recycling and recovery from wastewaters (e.g. Rosemarin et al. 2009; Cordell et al. 2009, 2011; Schröder et al. 2010; Pearce 2011; Walan 2013; Wyant et al. 2013; Yoshida et al. 2013; Heckenmüller et al. 2014; Desmidt et al. 2015).

However, as Steen (1998) pointed out, reserve/resource estimates are subjective as they depend on standards and criteria assumed by the data provider and determinations of the circumstances that might render a deposit economically useful. For example, Rosemarin et al. (2009) and Cordell et al. (2009) claimed that the rate of production of economically available phosphate reserves would peak between 2030 and 2040, after which demand would exceed supply, leading to global P scarcity. They also signalled that global phosphate reserves would start to run out within 75–100 years. Rosemarin et al. (2009) suggested that the most effective way to minimize the impacts of phosphate shortages would be to promote the recovery and reuse of P and other nutrients from organic waste and wastewater streams. They also called for policy reforms to promote the development of recycling technologies, in order to minimize EU dependence on P rock imports. Similarly, Shroeder et al. (2010) and Yoshida et al. (2013) highlighted that the full recycling of P would become essential for European and global security.

On the other hand, a number of scientists have pointed out that phosphate reserves may last between 300 and 380 years. For example, Walan (2013) investigated global P reserves forecasting and developed a model for peak P. He concluded that although today's estimated reserves based on the current production rates suggest the duration of over 300 years, a scarcity of P may occur much earlier than that. Moreover, he underlined that Chinese production could peak in the next 10–20 years which may have a considerable impact on the world production. Using the US Geological Survey data on global P production and reserves from 2010 to 2011 from (USGS 2012), Desmidt et al. (2015) estimated that the current phosphate rock reserves would be fully depleted in about 372 years. More recently, Daneshgar et al. (2018) stated that according to current research, the world reserves of the available P would not deplete before 2300.

Regardless from the discrepancies in forecasting phosphate reserves estimates, in the past 10 years, there has been a universal acknowledgement of the need for P recovery and recycling from wastewaters. This important recognition also leads to considerable branding and changes in the language used to describe crucial P issues and their implications on the environment, ecological and human health, and society. The term 'P pollution' has been branded to 'P – the critical resource', and term 'P removal' became 'P recovery' (e.g. Cordell et al. 2009; Rosemarin et al. 2009; Schroeder et al. 2010; Pearce 2011; Desmidt et al. 2015; UN Water 2017).

## 4.2 P Recovery from Municipal Wastewater Treatment Effluents

Mihelcic et al. (2011) estimated that 22% of global P demand could be satisfied by recycling human urine and faeces worldwide. However, Daneshgar et al. (2018) pointed out that only about 16% of mined P is consumed in a human diet, and that even if 100% recycling of P from human waste could be achieved, this would only reduce dependence on mined P by 16%. Nevertheless, similar to P removal from wastewaters (Section 3.2), to date, the majority of investigations on P recovery and re-use methods are focused on municipal wastewater treatment effluents (e.g. Wyant et al. 2013; Desmidt et al. 2015; Daneshgar et al. 2018). This may be due to the fact that approximately 90% of the wastewater P in MWWTPs is incorporated into the sewage sludge during the conventional chemical and biological P removal processes (Section 3.2).

Recovery of P can be implemented in different stages of treatment, from the liquid to the sludge phase, and from sludge post-treatment, such as incinerated sludge ash (Roeleveld et al. 2004; Cornel and Schaum 2009; Yoshida et al. 2013; Desmidt et al. 2015; Egle et al. 2016; Daneshgar et al. 2018). The majority of the available techniques for the recovery of P are based on the addition of calcium or magnesium salts, (and where needed seed crystals), to the P-rich wastewater and P recovery as calcium phosphate or struvite (e.g. Cornel and Schaum 2009; Yoshida et al. 2013; Desmidt et al. 2015; Daneshgar et al. 2018). P recovery as a struvite has been the major method, and it also dominated the scientific research presented at the Second European Sustainable Phosphorus Conference (ESPC2) held in Berlin in 2015 (Scope Newsletter 2015). The most commonly used processes for P recovery from sewage sludge and sewage sludge ash by are wet chemical or thermal processes. The wet chemical treatment of sewage sludge involves two main steps: (i) the addition of acid or base in order to dissolve P bound in sewage sludge; (ii) removal of insoluble compounds and separation of phosphates from the phosphorus-rich water, e.g. via precipitation, ion exchange, nanofiltration, or reactive liquid–liquid extraction (Cornel and Schaum 2009; Rapf et al. 2012; Yoshida et al. 2013). Cornel and Schaum (2009) pointed out that the same technologies can be applied to recover P from sewage sludge ash.

Rapf et al. (2012) underlined that thermal treatment of the sludge is limited only to regions that can afford this way of sludge disposal and therefore only a few thermo-chemical P-recovery processes have been developed in the past years. One of the early projects was the EU funded SUSAN-project whose aim was to develop a sustainable and safe strategy for nutrient recovery from sewage sludge using thermal treatment. The treatment consisted of two steps: (i) Mono-incineration of sewage sludge to eliminate pathogens and organic pollutants. However, although the incineration residues from this step have a high phosphorus content, it has low bioavailability. Moreover, the residues also contain heavy metal compounds above the limits for agricultural use. Therefore, the second step is necessary in order to remove heavy metals and transform P into mineral phases that would be available for plants; (ii) the thermochemical process is conducted where sewage sludge ashes are mixed with chlorine donors ($MgCl_2$ or KCl), grained, and treated in a furnace at 850–1000 °C to evaporate volatile heavy metal chlorides. Thus, the principle of the thermochemical method is the separation of the small heavy metals fraction from the sewage sludge ashes to obtain a P rich product that also contains $SiO_2$, CaO, $Al_2O_3$, $Fe_2O_3$, MgO, and $K_2O$ (e.g. Adam et al. 2009; Havukainen et al. 2016; Adam 2017).

Once dissolved, P could be recovered through adsorption or crystallization methods. One of the most frequently investigated and used P recovery technologies is a crystallization of magnesium ammonium phosphate (MAP), e.g. struvite ($NH_4MgPO_4 \cdot 6H_2O$). One of the major advantages is that recovered struvite contains less heavy metals, pathogens, and organic pollutants, and as such could be used directly on land as a slow-releasing fertilizer (Ueno and Fujii 2001; Yoshida et al. 2013). Typically, it contains 9.8% magnesium, 7.3% ammonium, 38.8% phosphate, and 44.1% water and other organic compounds. Total P recovery efficiency rates reported for struvite-based processes range from 50% to 80%. In addition, they can reduce ammonia concentrations in waste streams by 29% (Bird 2015).

Yoshida et al. (2013) provided a comprehensive summary of P recovery technologies from liquid waste streams (e.g. Struvite crystallization, Hydroxyapatite (HAP) crystallization, Electrolysis, Adsorption), P bound to sludge or slurry (Cambi-KREPRO, HeatPhos, AquaReci, Seabourne), and via incineration from ash and sludge (Bio-Con, Sephos and Advanced Sephos, and Phosphorus recovery from Ash [PASH]). They also provided brief description of these processes' main advantages and disadvantages and forms of recovered P. The most prevalent ones being calcium phosphates (HAP crystallization, HeatPhos, AquaReci, and Sephos), struvite (Struvite crystallization processes such as Ostara Pearl, AirPrex, PHOSNIX, Seabourne) and iron phosphate (Fuji Clean CRX, Cambi-KREPO). Some of the most popular commercial technologies are described later in the chapter.

In 2016, the Lippeverband, a public German water-management association and 10 other universities, public and private institutions from northern and western Europe (Germany, Belgium, France, Netherlands, Ireland, and Scotland) were awarded € 11.02 million for a four-year project focused on P recovery from municipal wastewater treatment effluents (Interreg 2019). The project website provides a sheet for seven P recovery technologies of which three have been specifically developed for P recovery from small scale sewage works (Table 4.1).

The European Sustainable Phosphorus Platform (ESPP) provided a comprehensive list of all EU and non-EU funded current and completed research and development projects focused on nutrient recycling and management. Currently, there are 132 EU funded projects and nearly as

**Table 4.1** European innovative phosphorus recovery technologies.

| Process | Product | Location and installation date |
|---|---|---|
| EuPhoRe®, thermal treatment of the sewage sludge via pyrolysis/ dry carbonization, followed by incineration | Phosphate ash directly ready for use as a fertilizer. | Germany, Dinslaken, outdoor, commissioned in December 2018. Input material: Sewage Sludge, 25–30% Dry Matter. Input mass: approx. 100 kg/h Output: Phosphate slag (12–20% $P_2O_5$) Output mass: approx. 10–15 kg/h. |
| The REMONDIS TetraPhos® (TetraPhos process) for the recovery of P as phosphoric acid from ashes originating from fluidized bed combustion of municipal sewage sludge. Over 80% of the ash P is recovered. | 1 RePacid®, a phosphoric acid ($H_3PO_4$) suitable for industrial applications or the use as raw material for further chemical processing; 2 The process also enables the recovery of Fe and Al salts as coagulants, for recycling. | Germany, Sewage sludge incineration plant WFA Elverlingsen GmbH, Werdohl Germany, indoor. Commissioned in May 2018. Input material: Sewage Sludge, approx. 50 kg DM/h Output mass: approx. 18 kg/h |
| STRUVIA™ Bio-acidification process combines the bio-acidification of sludge to solubilize phosphorus with the precipitation of struvite. The bioacidification is induced by adding easily degradable carbohydrate source in the sludge in strictly anaerobic conditions. | The struvite produced can be directly used as fertilizers. | Location: Lille (France) Commissioned: May 2018 Input material: Biological thickened sludge. Input mass: approx. 4 tons/day. Output mass: approx. 9–10 kg/day as Struvite (MAP) or 'phosphate salts' products. |
| PULSE Process involves a chemical acid leaching of P from partially to fully dried sewage sludge, followed by a reactive-extraction step and fractionated precipitation to remove contaminants. The last step is the precipitation of a fertilizer grade calcium and/or magnesium phosphate. | Phosphorous can be precipitated either as Ca or as Mg phosphate depending upon what is more desirable for the fertilizer industry. | The mobile demonstration plant will be installed at several different WWTP: Belgium/Tenneville, Ireland/Carrigrennan, Scotland/ Bo'ness, Germany/Dorsten. Commissioned in October 2019. Input material: partially/fully dried sewage sludge. Output mass: 1 kg/day of Ca/Mg/K phosphate |
| Small-Scale Sewage Treatment Works P recovery via Microalgae *Chlamydomonas acidophila*, which grows at a pH of 2–3, And has high phosphorus and nitrogen uptake rates (up to 90%) whilst operating at low temperatures. | After harvesting, the biomass of microalgae containing the recovered P and N can be used directly as fertilizers. The product could be spread on land either mixed with organic fertilizers or mixed with a liquid in the drip irrigation system. | Location: Scotland, Bo'ness Commissioning: September 2018 Input material: waste water at small scale WWTPs. Output mass: approx. 1–2 kg/week. |

**Table 4.1** (*Continued*)

| Process | Product | Location and installation date |
|---|---|---|
| The FiltraPHOS™ process is based on a rapid gravitational filtration of raw water through a granular media and its continuous self-backwashing. | After filtration, P rich sorbent material can be used directly as fertilizer or as an intermediate for the industry. | Location: Scotland, piloted by the Environmental Research Institute (ERI), part of University of the Highlands and Islands, and Véolia Commissioned: August 2018 Input material: waste water (low volume) Input mass: approx. 10 PE Output: Sorbent material enriched in PO4 Output mass: To be confirmed |
| Struvia™ process downscaled for small-scale WWTPs. It is based on an 'all-in-on' single stage crystallization reactor, which combines a Turbomix® draught tube impeller, a crystallization zone and a lamella separator in a single tank. | Expected products will be offered as 'phosphate salts' fertilizers. | Location: The mobile-set up of the Struvia reactors is run in Ireland and Scotland Commissioning: July 2018 Input material: WWTP streams Input mass: approx. 200 l/h pilot demonstrator |

Source: Interreg (2019).

many (117) completed projects. In addition, there are 41 current and 60 completed projects funded by the sources outside the EU (van Dijk 2018).

Ohtake and Tsuneda (2019) described 20 technologies for P recovery from sewage, sewage sludge incineration ash, steelmaking slag, manure, and bones. These included Struvite Recovery from Digested Sewage Sludge, Phosphorus Recovery from Sewage Sludge by High-Temperature Thermochemical Process (KUBOTA Process), Phosphorus Extraction from Sewage Sludge Ash by the $CO_2$ Blowing Method, Outotec (AshDec®) Process, the Ecophos Process, the Stuttgart Process, hydrothermal carbonization process. Most of these technologies were previously reviewed in Scope newsletters (Scope Newsletter 2018).

Schaum (2019) reviewed both P recovery and removal technologies. However, he focused only on one single P pollution source, the municipal wastewater treatment (sewage) plants. Technologies and processes reviewed include superparamagnetic particles, electrodialysis, thermal recovery, recovery by algae or bacteria, calcium phosphate precipitation, struvite recovery (Ostara Pearl and WASSTRIP, AirPrex, Phospaq, Phosnix), Mephrec slag, Remondis Tetraphos, Parforce, Leachphos, as well as those described by Ohtake and Tsuneda (e.g. Extraphos, Stuttgart, AshDec, Kubota). Only a few, the most popular processes, will be described.

### 4.2.1   Commercial Technologies and Products

#### 4.2.1.1   Phosnix Process

The Unitika Ltd Phosnix process was developed in Japan, 20 years ago. The process is based on the addition of magnesium to the side stream crystallization reactors where P is removed from the return liquor via the crystallization of magnesium ammonium phosphate hexahydrate ($NH_4MgPO_4 \cdot 6H_2O$). The principal idea behind the technology is to remove struvite ($MgNH_4PO_4 \cdot 6H_2O$, known also as MAP) from a wastewater treatment environment and then sell it to the fertilizer industry (Ueno and Fujii 2001), e.g. the same as the Ostara technologies products offered today.

#### 4.2.1.2   Ostara

Ostara Technologies, based in Vancouver, BC, Canada was established in 2005. Their process is very similar to Phosnix, and other P recovery technologies based on the addition of magnesium for controlled precipitation of struvite. However, thanks to the powerful promotion, skilful marketing, and branding (e.g. P recovery in crystallization reactors was branded into 'Pearl process', struvite precipitate into 'Crystal green' fertilizer) they became the leading and the most extensively promoted technologies both in North America and Europe (Cullen et al. 2013; *The Guardian* 2014; Ostara 2019a). Today there are 18 full scale Ostara plants implemented worldwide (Ostara 2019a) which is almost 50% of all large-scale plants implemented for P-recovery through struvite worldwide (Zhou 2017).

In the Pearl process, magnesium is added to P-rich waste streams (pre- and post-digestion thickening and dewatering liquors) to enable controlled precipitation of struvite (MAP). Crystalized nutrients resulting from the process are sold under the brand name Crystal Green. WASSTRIP process is designed to divert P from the sludge stream before it reaches the anaerobic digester, sending it directly to the Pearl reactor for optimal recovery. Detailed fact sheets and brochures about the Ostara Processes are available on their website (Ostara 2019a).

In 2013, Thames Water partnered with Ostara Nutrient Recovery Technologies to launch the UK's first full-scale nutrient recovery facility, at Slough sewage treatment plant, UK. The cost of the facility was £2 million, and it was estimated that the implementation of the Ostara Process would save around £200 000 per year in operating costs to Thames Water, by reducing chemical costs and preventing struvite deposits which could otherwise clog pipes and valves. However, it recovers only 10% of the total P loading into sewage treatment works (Scope newsletter 2013).

To date, there has been no information on whether this Ostara Process recuperated the capital investment or fulfilled Thames Water's expectations. In 2016, the Ostara Process was put in the operation at the Stickney Water Reclamation Plant located at the Metropolitan Water Reclamation District of Greater Chicago (MWRD), which is the largest wastewater treatment facility in the world (Ostara 2019b). Ostara is under obligation to make payments of US$ 400 for every ton of fertilizer recovered, for the next 20 years (Koch et al. 2015).

The Ostara process continues to be the most popular one worldwide, with 18 full scale operations across the globe. Their major selling points are that capital investments are recovered through savings in operating costs (e.g. maintenance, metal salt addition, sludge disposal, $CO_2$ credits, and so on). Although the Ostara process has high operating costs ($MgCl_2$ addition, NaOH addition, power, labour, maintenance), they claim that these costs can be recovered through Crystal Green Revenue within 3–10 years (Koch et al. 2015).

### 4.2.1.3 Stuttgart Process

The 'Stuttgart Process' is another process based on struvite precipitation, developed at the University of Stuttgart, Germany. The first large scale plant was put into operation in autumn 2011 (Antakyali et al. 2013). Although this first project was considered pioneering in Germany and gained considerable interest by the authorities and public, it revealed that the process was not economically viable compared to the cost of fertilizer produced from mined P. Egle et al. (2016) also pointed out that recovering P from sewage sludge is generally more expensive than recovering P from supernatant. They estimated that the cost of 1 kg of recovered P produced by the Stuttgart process ranges from 9 to 16 € depending on the chemicals used. Amann et al. (2018) highlighted that all technologies recovering from sewage sludge show a considerable increase in cumulative energy demand (CED) compared to the reference system.

### 4.2.1.4 NuReSys

Nutrients Recovery Systems (NuReSys) is another technology based on controlled struvite crystallization, developed in Belgium. The key offering is that the process can be applied both on digested sludge or post dewatering, as well as in a combination of approaches. Several systems have been implemented at full industrial scale. Details about technology and applications are provided at NuReSys website (NuReSys 2019). Sarvajayakesavalu et al. (2018) recently stated that the operating expenditures (OPEX) for the NuReSys process treating $60 \, m^3/h$ wastewater, containing $120 \, mg/l \, PO_4 - P$ is 1.6 EUR/kg P, whilst the capital expenditures (CAPEX) is 4.4 EUR/kg P. They pointed out that although P recovery is considered to be viable, environmentally safe, and technically feasible, its economic feasibility is fairly limited. This is the case with all other P recovery technologies currently available on the market (Prabesh 2018).

### 4.2.1.5 Outotec (AshDec) Process

The Outotec ASH DEC technology employs a thermochemical process for separating heavy metals from ash generated during the combustion of industrial and municipal waste. P is than recovered from ashes generated during combustion. The technology was originally developed by ASH DEC

Umwelt AG, a company based in Austria. The Outotec acquired technology in 2011 (Outotech News 2011).

Following a 50 million euro investment, in the summer of 2015, Outotec completed the largest sewage sludge incineration plant in Switzerland, providing treatment for and handling all the sludge produced in the Zürich canton area, amounting to 100 000 metric tons a year (Outotech News 2016).

## 4.3  P Recovery from Manure

The potential for P recovery from animal manure via struvite precipitation has been documented for nearly 30 years (e.g. Wrigley et al. 1992; Burns and Moody 2002; Schoumans et al. 2010). Recent data from the literature indicate that 90–95% of P can be recovered from manure through struvite or calcium containing P precipitate (Abella et al. 2014; Szögi et al. 2015; Tarayre et al. 2016).

Several other methods and technologies were developed in the past 10–15 years such as Solid–Liquid Separation, Electrolysis, and Thermochemical Conversion (pyrolysis, combustion, gasification, hydrothermal carbonization) (e.g. Desmidt et al. 2015; Schoumans et al. 2015). Microalgae cultivation had been proposed as another potential method of nutrient extraction (Foged et al. 2011). Vanotti et al. (2005) developed and patented (US patent 6,893,567) a process to recover phosphate from liquid swine manure. In their treatment system polymers are added to the raw liquid swine manure and treated in an enhanced solid–liquid separation process; the liquid swine manure is then treated with nitrification to oxidize ammonium to nitrate (Desmidt et al. 2015).

However, a number of technologies applied on farms at the full scale are very limited. One of the reasons may be in the challenges associated with the high content of organic compounds present both in manure and anaerobic digester effluents (e.g. Desmidt et al. 2015; Schoumans et al. 2015; Tarayre et al. 2016). Another reason may be in the fact that similar to P removal technologies, the costs of the P recovery process installation and operation on farms cannot be recovered via the same mechanisms used for MWTP's upgrades and installations, e.g. through water tariffs, or a mix of tariffs, transfers, and taxes (EurEau 2019), because such a funding mechanism does not exist for agricultural wastewater treatment. Therefore, it is much harder to sell and/or ensure return on investment if attempting to promote and offer P recovery technologies in this market, as funding sources would have to come directly from farmers, e.g. private sources. Moreover, as the cost of P recovered from manure is much higher compared to the mineral fertilizer, there are no economic incentives for farmers to invest in P recovery processes and technologies on their livestock operations. This situation creates a considerable gap in research and development of novel processes and technologies to achieve more cost-effective nutrient recovery from manure.

For example, of the 140 projects funded by the major EU (H2020 [FP], LIFE, INTERREG) and national/industries programmes, and completed between 2011 and 2018, only 10 focused on processes and/or technologies for P (or both P and N) recovery from manure (Table 4.2).

Only a few of above listed projects resulted in field applications of P (and other nutrients) recovery technologies/processes. For example, as a part of DeBugger project (3) in June 2015, a pilot process was implemented at the wastewater treatment plant in Sweden (Ends waste and bioenergy 2015;

**Table 4.2** List of EU projects investigating technologies and processes for P recovery from manure.

| | Project Name | Duration | Source |
|---|---|---|---|
| 1 | Baltic Forum for Innovative Technologies for Sustainable Manure Management (BALTIC MANURE) led by the Natural Resources Institute, Finland | 2011–2013 | http://eu.baltic.net/Project_Database.5308.html?contentid=58&contentaction=single |
| 2 | Process for sustainable phosphorus recovery from agricultural residues by enzymatic process to enable a service business for the benefit of European farm community (PhosFarm); led by Fraunhofer Institute for Interfacial Engineering and Biotechnology IGB, Germany. | 9/2013–9/2015 | https://www.phosfarm.eu/project.html |
| 3 | Demonstration of efficient Biomass Use for Generation of Green Energy and Recovery of Nutrient (DeBugger); lead by Outotech AB company, the Netherlands. | 1/2013–12/2015 | http://www.innoenergy.com/case-study/debugger |
| 4 | Nutrient recovery from manure (ReuseWaste), led by the University of Copenhagen, Denmark. | 1/2012–12/2015 | www.reusewaste.eu |
| 5 | New technological applications for wet biomass waste stream products (NEWAPP) awarded to the European Biomass Industry Association. | 11/2013–4/2016 | http://www.eubia.org/cms/projects-2/completed-projects/newapp-2 |
| 6 | Green fertilizer upcycling from manure: Technological, economic and environmental sustainability demonstration (ManureEcoMine) led by Gent University, Belgium. | 11/2013–10/2016 | http://www.manureecomine.ugent.be/0 |
| 7 | An innovative bio-economy solution to valorise livestock manure into a range of stabilized soil-improving materials for environmental sustainability and economic benefit for European agriculture (BioEcoSIM) project led by Fraunhofer IGB, Germany. | 10/2012–12/2016 | www.bioecosim.eu |
| 8 | Phosphorus Recycling of Mixed Substances (BONUS PROMISE), also led by the Natural Resources Institute Finland. | 4/2014–3/2017 | https://www.bonusportal.org/projects/innovation_2014-2017/promise |
| 9 | Swine-farm revolution (DEPURGAN) led by the Spanish company Eurogan. | 9/2015–7/2017 | http://www.depurgan.com/ingles/project-depuration.php |
| 10 | Recycling inorganic chemicals from agro- and bio-industry waste streams (Biorefine project,) funded by the Interreg IVB North West Europe (Schoumans 2014; Schoumans et al. 2014), and led by Ghent University. | 2011–2015 | https://www.ugent.be/bw/agricultural-economics/en/research/moderna/past-projects/biorefine.htm |

Source: van Dijk (2018).

Outotech 2019a). The DeBugger system was promoted as a closed-loop steam dryer capable of saving up to 50% of the energy needed to evaporate water contained in biomass compared to anaerobic digestion plants. The system also provided thermal treatment of the dried substrate. However, there is no information on the process performance, nor whether it is still in operation (Outotech 2019a). Since the completion of DeBugger project in 2015, Outotec company was awarded three additional projects (one as the lead investigator, two as co-investigator): (i) Closing the Global Nutrient Loop (CLOOP), 11/2017–10/2020, which aims to tackle the current luck of demand for recycled nutrients by creation of a different perception of fertilizer quality (Outotech 2019b); (ii) Sustainable Management of Phosphorus in Baltic countries (InPhos), 1/2018–21/2019, which is focused on the knowledge transfer and design of solutions for the sustainable use of P; and (iii) Nutrient Recycling – from pilot production to farms and fields (ReNu2Farm), 9/2017–8/2020 (Table 4.3).

The technology for manure treatment developed during BioEcoSIM project (7) was introduced to the agricultural wastewater treatment market in spring 2018 and is currently offered by SUEZ Germany as an operator of large-scale plants (Fraunhofer-Gesellschaft 2018). A technology for swine manure treatment developed during DEPURGAN project (9) is currently offered by a company from Zaragosa, Depurgan (Depurgan 2019).

Following the completion of Biorefine, a much larger project was established, Biorefine Cluster Europe (2015 to date). The aim of this Cluster project is to interconnect projects and people within the domain of biobased resource recovery and enable more sustainable resource management. The central research theme of this project is in the biorefinery sector, e.g. the refinement of chemicals, materials, energy, and products from biobased waste streams. Resource recovery and extraction from biomass represent one of the project components along with bioprocesses, bio-energy production, and biobased waste streams (https://www.biorefine.eu/about).

### 4.3.1   Current EU P Recovery Projects

Currently, there are 132 nutrient research and development projects funded by the EU. Of these, only two are focused on the nutrient recovery from manure: (i) *Large scale demonstration projects for recovery of nutrients from manure, sewage sludge and food waste* (SYSTEMIC) project, led by Professor Schoumans from the Wageningen University & Research (WUR), 6/2017–6/2021, investigates the new approaches for the valorisation of biowaste into green energy, mineral fertilizers, and organic soil improvers (https://systemicproject.eu/about/#conso). The installation of five demonstration-scale nutrient recovery plants is planned, operating in combination with large anaerobic digesters and field testing of the recovered nutrient fertilizer products from manure, sewage sludge, and food waste to demonstrate agronomic value, business case, and environmental benefits; (ii) *Duckweed*

**Table 4.3** Current EU projects focused on P recovery from manure funded by non-EU sources.

| Project Name and Description | Source |
|---|---|
| 1  *The Added Value from Manure and Minerals (Meerwaarde Mest en Mineralen)* is funded by the Dutch Government, is led by Professor Schoumans at WUR. It investigates integrated manure processing from an anaerobic digestion process and its potential for scaling up across several regions. The added value in the chain involves investigations of potential buyers of products, the builders of installations (manufacturing industry), governments (national and provincial authorities), and water boards. The approach is based on the methodology and strategy that was developed in the first phase of the Added Value for Manure and Minerals Public Private Partnerships which investigated the potential for P recovery on a laboratory and small scale pilot scale. This previous project led to the establishment of a Green Mineral Plant where not only phosphate but also nitrogen is recovered, and organic matter with reduced nitrogen (N) and phosphorus (P) content remains available for Dutch food production | van Dijk (2018); https://www.wur.nl/nl/show/Meerwaarde-mest-en-mineralen-AF12178.htm |
| 2  *Wetsus Phosphate recovery from iron phosphate and iron-based phosphate adsorbents* are led by Wetsus, the European centre of excellence for sustainable water technology located at Leeuwarden WaterCampus, the Netherlands. Its major focus is on the development of new technologies for both sewage wastewater and manure treatment that would provide both P removal and P recovery as concentrated high value products. Wetsus also maintains the inventory of nutrient recovery technology reviews. | Wetsus (2019); https://www.wetsus.nl/phosphate-recovery |
| 3  *Biobased Fertilizers (Achterhoek/Kunstmestvrije Achterhoek)* is funded by the INTERREG Germany-Netherlands, Ministry of Agriculture, Nature and Food quality, and Province of Gelderland, Netherlands consists of four related tasks: (i) Production of high-quality biobased fertilizers (BBF), (ii) Distribution and application of the BBF, (iii) Investigations of the BBF agricultural value and potential negative impacts on the environment, and (iv) Provision of advice services for customers and market development. In addition, the outcome of this project will provide reliable data and vital information for policy discussions and development in Europe. | van Dijk (2018); www.kunstmestvrijeachterhoek.nl |
| 4  *Nutrient Recycling – from pilot production to farms and fields* (ReNu2Farm), is led by IZES gGmbH, from Germany, and involves partners from five additional countries in North West Europe (NWE) (UK, Ireland, Netherlands, Belgium, and France). The major aim of the project is to enable nutrient exchange between partner countries. The main research plans are to (i) investigate the current situation on nutrients and technologies for their recovery in NWE region; (ii) deploy some of these technologies in practice, for both the production from sewage sludge, food wastes and manure, and for recycled nutrients products upcycling; (iii) identify possible market barriers; (iv) conduct an assessment of the regional demand for nutrients and performances of the products; and (v) identify legal requirements and pressure points. | http://www.nweurope.eu/projects/project-search/renu2farm-nutrient-recycling-from-pilot-production-to-farms-and-fields |

*technology for improving nutrient management and resource efficiency in pig production systems* (Lemna) led by the Ainia technological centre, Spain, 10/2016–12/2019. The main aim of this project is to implement and assess the performance of a duckweed pilot scale ($3\,m^3$/day) production system in sustainable treatment and nutrients recovery from the anaerobically digested pig manure (http://www.life-lemna.eu).

Of the 41 projects funded by the sources outside the EU, only four focus on P recovery processes from manure (Table 4.3). Three of these projects are led by the institutions in the Netherlands, the leading country in promoting circular agriculture. Following the proposal from WUR, in autumn 2018 the Dutch government committed to develop a fertilizer-free livestock farming by 2020 and create a balance on the manure market. The core of their strategy is to diminish the supply of P to agriculture through animal feed and fertilizer, optimize the use of phosphate and animal manure within the Netherlands, and to increase the export possibilities of phosphate through phosphate or P recovery (WUR 2018; van Dijk 2018).

Several of the above projects highlight the need to develop strategies and create demand for recycled nutrients. Apart from CLOOP led by Outotech (Outotech 2019b, Section 4.3), the Added Value from Manure and Minerals (Table 4.3, item 1) project investigates ways to reduce the supply of phosphate to agriculture through animal feed and fertilizer, to optimize the use of phosphate and animal manure within the Netherlands, and increase the export possibilities of recovered phosphate and P (van Dijk 2018). Building on the need to increase the use of recycling-derived fertilizer products by farmers, one of the aims of the ReNu2Farm project (Table 4.3, item 4) is to investigate market barriers for wider use of recycled fertilizer products amongst six countries in North West Europe.

## 4.4 P Recovery from Alternative Sources – Water and Soil Management Systems

Although the majority of the investigations of P recovery and recycling from wastewater effluents to date have been on municipal sewage effluents at WWTPs and animal manure, several researchers explored less obvious, alternative sources, such are P-removing filters, bioretention and other management practices for urban stormwater runoff treatment (Section 3.6).

As the entire research motivation behind the exploration of industrial-by products as potential P filtration media 20 years ago (Section 3.3.3.2) was in finding the appropriate after-use of the waste materials, closing the loop and reducing the quantities of waste materials, this research was also an enabler of resource recovery and circular economy. The first step in assessing the potential of filter materials for reuse as a P-rich soil amendment is to determine the plant availability of P retained within the media (Johansson and Hylander 1998; Hylander et al. 2006; Bird and Drizo 2009; Cucarella 2009). The second step is to perform leachate studies to ensure there is no leaching from the media that could cause adverse environmental affects (e.g. Bird and Drizo 2009). Some of the investigated materials included Ca rich soils, opoka, blast furnace steel slag aggregates (Johansson and Hylander 1998; Hylander and Siman 2001; Hylander et al. 2006), electric furnace steel slag aggregates (Bird and Drizo 2009), polonite, wollastonite, filtra P (Cucarella 2009). The results from this research revealed that all of the investigated materials were free from pathogens (due to high Ca content) and have a potential to act as a slow release P fertilizer (Yoshida et al. 2013).

However, the application of decentralized residential, agricultural, and/or urban stormwater runoff systems for P removal is severely hindered by the continuing absence of regulatory requirements for P removal from these sources and complexities associated with technology validation and permitting process (Chapters 2 and 3). Therefore, despite the obvious environmental, economic, and social benefits that reusing the spent filtration media as a slow P fertilizer could provide, it continued to be disposed of into landfills (Yoshida et al. 2013; Drizo and Picard 2015; Drizo et al. 2016; Drizo 2017).

Roy (2017) reviewed and assessed the potential for P recovery and recycling from several alternative onsite decentralized systems used in ecological engineering, e.g. composting, vermicomposting (composting with earthworms), biogas residues resulting from the anaerobic digestion process, assimilation of P by vegetation and algae (e.g. bioretention systems, ponds, constructed wetlands), phytoextraction (using vegetation to extract excess P from soil or water), zooextraction (concentrate P in animal biomass to harvest P), aquaculture, P-removing filters, biochar (organic material product of pyrolysis, a process by which material is heated to temperatures of 300 °C, under oxygen-limited conditions.

Building on the transdisciplinary framework for implementing a regenerative urban P cycle, a Phosphorus Recovery Transition Tool (PRTT) developed by Pearce (2015), Roy (2017) discussed potential drivers and barriers of P recovery and recycling strategies from ecological engineering systems. He highlighted that despite considerable research showing the potential for P recovery from these alternative systems, there is a large number of economic (e.g. lack of market demand for recovered P; operational and labour costs), environmental and health (heavy metals leaching, pathogens content), technical (scaling, infrastructure), regulatory (e.g. absence of standards, nutrient limits, regulatory requirements, public misperceptions), organizational (e.g. lack of knowledge of new protocols, technology, and products logistics) new protocols, and individual behaviour barriers (perception towards new technology, new habits).

## 4.5 Phosphorus Recovery Regulations

Unlike with P removal from wastewaters, the global recognition of the need for P recovery coupled with several full scale installations in Europe (the Netherlands, Belgium, Germany, Austria), North America, and Japan instigated the development of the new regulatory framework for P recycling and reuse in several countries (Desmidt et al. 2015; Prabesh 2018).

This has also been supported by the growing legislation imposing restrictions and even the prohibition of the agricultural use of sewage sludge (Wyant et al. 2013; Inglezakis et al. 2014; Desmidt et al. 2015; Zhou 2017). These restrictions will require new methods and practices for sludge disposal and treatment and could bring about a considerable burden to the wastewater treatment industry (e.g. Kelessidis and Stasinakis 2012; Inglezakis et al. 2014; Prabesh 2018). Treatment and final disposal of sewage sludge in European countries have been recently described by Kelessidis and Stasinakis (2012), whilst Inglezakis et al. (2014) provided a comprehensive review of the EU legislation related to sewage sludge management. In the majority of Member States, the regulations specific to the disposal and recycling of sludge are focused on the use of sludge in agriculture; sludge disposal is addressed by general legislation on landfill and incineration of waste.

Desmidt et al. (2015) discussed the influence of legislation and national policies on the implementation of P-recovery technologies. He highlighted the fact that the implementation of P-recovery technologies was strongly influenced by national regulations and/or the presence of industrial facilities that can use recovered P. For example, the Netherlands has one of the most severe regulations governing the maximum heavy metal content of the sewage sludge for agricultural land application. Consequently, the Netherlands is one of the pioneers of the developing of P-recovery technologies and were awarded the largest number of research and development projects in the past 10 years (Section 4.3). However, P requirements in the soils are low and largely met by spreading animal waste and mineral fertilizers. Therefore there is no demand for recovered P as fertilizer (Desmidt et al. 2015; Sarvajayakesavalu et al. 2018). Sweden also has stringent national requirements on the spreading of sludge to agriculture land. Nevertheless, the Swedish Environmental Protection Agency (SEPA) was amongst the first one in Europe to propose a target for P-recycling that by 2015, at least 60% of the phosphorus in wastewater should be restored to productive soil, of which half should be returned to arable land (Stark 2007; cited by Desmidt et al. 2015 and Prabesh 2018). Switzerland and Germany also introduced regulatory frameworks relating to the recycling and recovery of P. In Switzerland, the Ordinance on Avoidance and Disposal of Waste (VVEA) requires the recovery of P from wastewater, sewage sludge, and sewage sludge ashes and the material utilization of P in meat and bone meal as of 2026. In 2017, Germany implemented the amendment of the national Sewage Sludge Ordinance which obliges wastewater treatment plants bigger than 50 000 population equivalents to implement P recovery within the next 12–15 years (Mehr et al. 2018). However, similar to P-removal technologies, the new regulatory framework for P recovery is mainly focused on municipal wastewater, sewage sludge, and sewage sludge ashes, whilst other sources (agricultural manure, effluents and runoff, residential septic systems and/or urban stormwater runoff) are excluded. Major Rules and Regulations are summarized in Table 4.4.

Table 4.4 Summary of the major EU rules and regulations related to P recovery, recycling, and reuse.

| Date | Proposal/Rule/Legislation | Source |
|------|---------------------------|--------|
| 26/5/2014 | Phosphate rock added to the list of Critical Raw Materials for which supply security is at risk | European Commission (2014) |
| 2/12/2015 | Circular Economy Package | European Commission (2015d) |

*Purpose*: a new strategy to boost global competitiveness, foster sustainable economic growth, and generate new jobs within the Member States. The Package contained aspiring Action plan which included several crucial points to stimulate P recovery, recycling, and re-use. The most important legislative measure proposed was 'a revised Regulation on fertilisers, to facilitate the recognition of organic and waste-based fertilisers in the single market and support the role of bio-nutrients'.

| | | |
|------|---------------------------|--------|
| 17/3/2016 | New Regulations on organic and waste-based fertilizers in the EU | European Commission (2016b) |

*Purpose*: to enhance the use of organic and waste-based fertilizers, encourage their access to the EU Single Market and enable their offerings along with traditional, non-organic fertilizers. Also, the rules aimed to reduce the need for mineral-based fertilizers, and consequently dependence of Member States on imports of phosphate rock.

| | | |
|------|---------------------------|--------|
| 22/5/2018 | New Rules for EU waste legislation approved | European Commission (2018c) |

The new Rules stipulate new recycling targets for municipal waste, requiring 55%, 60%, and 65% of municipal waste to be recycled by 2025, 2030, and 2035, respectively.

| | | |
|------|---------------------------|--------|
| 12/12/2018 | Agreement on Commission proposal to boost the use of organic and waste-based fertilizers reached | European Commission (2018b), Biorefine (2018) |

*Purpose*: It is expected that the new rules will assist in generating new market opportunities for companies producing organic fertilizers, provide common rules on safety, quality, and labelling requirements across Europe, and concomitantly, assist in the creation of a new market for secondary raw materials in line with the circular economy.

## 4.6 Conclusions

The role of P as a principal trigger of eutrophication and major water quality impairment has been recognized for over 30 years. Yet, in the past three decades, the EU developed only a few regulations to address the issue (Table 2.3). Moreover, as highlighted throughout Chapters 2 and 3, current regulations only address a single P loading source, the municipal sewage wastewaters. The absence of regulatory requirements has imposed vast hindrances in the development of innovative P-removal technologies and their placement on the wastewater treatment market (Chapter 3).

On the other hand, although depletion of global P reserves, the increase in fertilizer prices, and the general urge for P recovery have received the attention from the scientific community and general public for only 10 years, the EU made an unprecedented progress in development of rules and regulations in the past five years to encourage P recovery and recycling from wastewaters. Unfortunately, similar to P removal from wastewaters, P recovery is most frequently encouraged only from municipal sewage treatment plants (Section 4.2).

Notwithstanding the obvious difference in the final products of traditional P removal and P recovery processes, if one attempts to mitigate eutrophication problems by preventing/minimizing the input of P at the source, the process will always start with P removal and/or harvesting. Regardless from whether one aims to obtain a P-free effluent ('P removal'), or to generate a P-containing product that can be reused either in agriculture or in P-industry ('P recovery') one has to remove/harvest P from the wastewater stream. Therefore all P recovery technologies need to have some sort of P-removal device, to begin with.

Despite remarkable scientific, industrial, and legislative support and encouragement for P-recovery research and development projects and creation of demand for P-recovery products, in particular within the EU member states, to date, there have been only a few full scale implementations, and mainly on sewage treatment plants (Section 4.2). Therefore, it is clear that even the substantial progress made in the past five years in funding and development of legislation to encourage P recovery and recycling – there has been no progress in reducing the extent of eutrophication and number of areas affected by HABs worldwide.

Although the critical resource, one has to keep in mind that P remains the principal trigger of eutrophication and HABs. Whilst very helpful, new P-recovery legislation (Table 4.4) is not sufficient to initiate development of a new generation of P-removal and harvesting technologies from agricultural, residential, or urban stormwater runoff. If we continue to neglect P loading from these waste streams, there will be no reduction in eutrophication. Moreover, Global Climate Change will promote cyanobacterial growth and exacerbate HABs at much larger scales. It will continue to diminish potable water supplies.

# References

Abegglen, C., Ospelt, M., and Siegrist, H. (2008). Biological nutrient removal in a small-scale MBR treating household wastewater. *Water Research* 42 (1–2): 338–346.

Abella, E.T., Puig, S., Balaguer, M. et al. (2014). Phosphorus recovery from pig/cow manure: a sustainable approach. SCOPE Newsletter (special edition no. 106): 100. https://phosphorusplatform.eu/links-and-resources/downloads (accessed 10 May 2015).

Adam, C. (2017). SUSAN – sustainable and safe re-use of municipal sewage sludge for nutrient recovery. https://opus4.kobv.de/opus4-bam/frontdoor/index/index/docId/39160 (accessed 27 April 2019).

Adam, C., Peplinski, B., Michaelis, M. et al. (2009). Thermochemical treatment of sewage sludge ashes for phosphorus recovery. *Waste Management* 29: 1122–1128.

Adam, K., Krogstad, T., Suliman, F.R.D. et al. (2005). Phosphorous sorption by Filtralite-P ™– smallscale box experiment. *Journal of Environmental Science and Health. Part A, Toxic/Hazardous Substances & Environmental Engineering* 40 (6–7): 1239–1250.

Adam, K., Krogstad, T., Vrale, L. et al. (2007). Phosphorus retention in the filter materials shells and filtralite P?—Batch and column experiment with synthetic P solution and secondary wastewater. *Ecological Engineering* 29: 200–208.

Adera, S., Drizo, A., Twohig, E. et al. (2018). Improving performance of treatment wetlands: evaluation of supplemental aeration, varying flow direction, and phosphorus removing filters. *Water, Air and Soil Pollution* 229 (3): 100. https://doi.org/10.1007/s11270-018-3723-3.

Agriculture and Agri-Food Canada (2004). Beneficial management practices. In: *Environmental Manual for Crop Producers in Alberta*, 175. Edmonton: Alberta Agriculture, Food and Rural Development http://www1.agric.gov.ab.ca/$department/deptdocs.nsf/all/agdex9483/$file/100_25-1.pdf?OpenElement (accessed 27 April 2019).

Agrifood Asia (2018). Industry Sectors Livestock China. http://www.agrifoodasia.com/English/ind_sectors/livestock.htm (accessed 7 November 2018).

Ahiablame, L.M., Engel, B.A., and Chaubey, I. (2012). Effectiveness of low impact development practices: literature review and suggestions for future research. *Water, Air and Soil Pollution* 223: 4253–4273. https://doi.org/10.1007/s11270-012-1189-2.

*Phosphorus Pollution Control - Policies and Strategies*, First Edition. Aleksandra Drizo.
© 2020 John Wiley & Sons Ltd. Published 2020 by John Wiley & Sons Ltd.

Alexander, R.B., Smith, R.A., Schwarz, G.E. et al. (2008). Differences in phosphorus and nitrogen delivery to the Gulf of Mexico from the Mississippi River Basin. *Environmental Science and Technology* 42: 822–830. https://doi.org/10.1021/es0716103.

Almeelbi, T. and Bezbaruah, A.N. (2012). Aqueous phosphate removal using nanoscale zero-valent iron. *Journal of Nanoparticle Research* 14 (7): 1–14.

Al-Zboon, K.K. (2017). Phosphate removal by activated carbon–silica nanoparticles composite, kaolin, and olive cake. *Environment, Development and Sustainability* 20 (6): 2707–2724. https://doi.org/10.1007/s10668-017-0012-z.

Amann, A., Zoboli, O., Krampe, J. et al. (2018). Environmental impacts of phosphorus recovery from municipal wastewater. *Resources, Conservation & Recycling* 130: 127–139. https://doi.org/10.1016/j.resconrec.2017.11.002.

American Society of Agricultural Engineers (2005). ASAE D384.2 MAR2005. Manure production and characteristics. http://www.agronext.iastate.edu/immag/pubs/manure-prod-char-d384-2.pdf (accessed 12 May 2015).

American Society of Civil Engineers, the ASCE (2017). 2017 Infrastructure report card. https://www.infrastructurereportcard.org/cat-item/wastewater (accessed 27 April 2019).

Amery, F. and Schoumans, O. (2014). Agricultural phosphorus legislation in Europe. https://www.ilvo.vlaanderen.be/portals/68/documents/mediatheek/phosphorus_legislation_europe.pdf (accessed 27 April 2019).

AncientPages (2016). First bathrooms appeared around 8,000 BC In Scotland. http://www.ancientpages.com/2016/12/11/first-bathrooms-appeared-around-8000-b-c-in-scotland (accessed 27 April 2019).

Anderson, D.L., Tuovinen, O.H., Faber, A. et al. (1995). Use of soil amendments to reduce soluble phosphorus in dairy soils. *Ecological Engineering* 5: 229–246.

Andersson, M.T. (2012). Paving the way to Asia for innovative environmental technologies. https://www.dhigroup.com/global/news/imported/2012/2/2/pavingthewaytoasiaforinnovativeenvironmentaltechnologies (accessed 27 April 2009).

Ansari, A.A., Sarvajeet, S.G., Lanza, G.R. et al. (eds.) (2011). *Eutrophication: Causes, Consequences and Control*. eBook. Dordrecht: Springer Publishing. ISBN: 978-90-481-9625-8.

Antakyali, D., Meyer, C., Preyl, V. et al. (2013). Large-scale application of nutrient recovery from digested sludge as struvite. *Water Practice & Technology* 8 (2): 256–262.

AquaEnviro (2016). Phosphorus removal. European Wastewater Management Conference, Manchester, UK, (11–13 October 2016). https://www.aquaenviro.co.uk/wp-content/uploads/2015/06/10th-EWWM-FINAL-Timetable.pdf (accessed 27 April 2019).

AquaEnviro (2018). Phosphorus removal. European Wastewater Management Conference, Manchester, UK, (17–18 July 2018). https://www.aquaenviro.co.uk/wp-content/uploads/2015/06/European-Waste-Water-Management-Conference-Final-Timetable-V5.pdf (accessed 27 April 2019).

Ashford, N.A. and Hall, R.P. (2011). The importance of regulation-induced innovation for sustainable development. *Sustainability* 3: 270–292. https://doi.org/10.3390/su3010270.

Baker, M.J., Blowes, D.W., and Ptacek, C.J. (1998). Laboratory development of permeable reactive mixtures for the removal of phosphorus from onsite wastewater disposal systems. *Environmental Science and Technology* 32 (15): 2308–2316.

Balmer, P. (2004). Phosphorus recovery – an overview of potentials and possibilities. *Water Science and Technology* 49 (10): 185–190.

Banack, S.A., Caller, T.A., and Stommel, E.W. (2010). The cyanobacteria derived toxin beta-N-methylamino-L-alanine and amyotrophic lateral sclerosis. *Toxins* 2: 2837–2850.

Barca, C. (2012). Steel slag filters to upgrade phosphorus removal in small wastewater treatment plants. PhD thesis. Chemical and Process Engineering. Ecole des Mines de Nantes, France.

Barnard, J.L. (1975). Biological nutrient removal without the addition of chemicals. *Water Research* 9 (5-6): 485–490.

Bashar, R., Gungor, K., Karthikeyan, K.G. et al. (2018). Cost effectiveness of phosphorus removal processes in municipal wastewater treatment. *Chemosphere* 197: 280–290. https://doi.org/10.1016/j.chemosphere.2017.12.169.

Beal, C.D., Gardner, E.A., and Menzies, N.W. (2005). Process, performance, and pollution potential: a review of septic tank-soil absorption systems. *Australian Journal of Soil Research* 43 (7): 781–802.

Behbahani, M., Moghaddam, A.M.R., and Arami, M. (2011). A comparison between aluminum and iron electrodes on removal of phosphate from aqueous solutions by electrocoagulation process. *International Journal of Environment* 5 (2): 403–412.

Bennett, O., Watson, C., and Camp, J. (2014). Septic tanks: new regulations. Standard Note: SN06059. 18 December, 2014. http://researchbriefings.files.parliament.uk/documents/SN06059/SN06059.pdf (accessed 27 April 2019).

Binev, D. (2015). Continuous fluidized bed crystallization. PhD thesis. Universität Magdeburg, Germany. 172 pp. http://pubman.mpdl.mpg.de/pubman/item/escidoc:2168043/component/escidoc:2227684/2168043.pdf (accessed 27 April 2019).

Biorefine (2018). Biorefine cluster bulletin. Policy News. https://www.biorefine.eu/sites/default/files/publication-uploads/mailchi-mp-2be730c48fe2-biorefine-bulletin-december.pdf (accessed 27 April 2019).

Bird, A.R. (2015). Evaluation of the feasibility of struvite precipitation from domestic wastewater as an alternative phosphorus fertilizer resource. Master's thesis project. University of San Francisco USF Scholarship Repository. http://citeseerx.ist.psu.edu/viewdoc/download?doi=10.1.1.1027.2229&rep=rep1&type=pdf (accessed 27 April 2019).

Bird, S. and Drizo, A. (2009). Investigations on phosphorus recovery and reuse as soil amendment from electric arc furnace slag filters. *Journal of Environmental Science and Health. Part A, Toxic/Hazardous Substances & Environmental Engineering* 44 (13): 1476–1483.

Bird, S. and Drizo, A. (2010). EAF steel slag filters for phosphorus removal from milk parlor effluent: the effects of solids loading, alternate feeding regimes and in-series design. *Water* 2 (3): 484–499. https://doi.org/10.3390/w2030484.

Blowes, D.W., Ptacek, C.J., and Baker, M.J. (1996). Treatment of wastewater. G.B. Patent 2,306,954 issued 1 Dec. 1999, 1996; Canadian Patent 2,190,933, filed 11 November 1996; U.S. Patent 5,876,606, issued 9 March 1999, Phosphex™ Canadian trademark registration 1,051,185 filed 17 March 2000, Phosphex™ US trademark registration 78/015,068 filed 30 June 2000.

Bloxham, P. (1992). The management of silage effluent. Technical paper, Harper Adams, Agricultural College, Farm Buildings and Engineering (9) (1): 21–23.

Bluewater Bio (2018). Filterclear – high rate multi-media filtration. http://www.bluewaterbio.com/filterclear/ (accessed 11 November 2018).

BMP Database (2018). International Stormwater BMP Database. http://www.bmpdatabase.org/history.html (accessed 27 April 2019).

Boehler, M., Joss, A., Buetzer, S. et al. (2007). Treatment of toilet wastewater for reuse in a membrane bioreactor. *Water Science and Technology* 56 (5): 63–70.

Boeuf, B., Fritsch, O., and Martin-Ortega, J. (2016). Undermining European environmental policy goals? The EU Water Framework Directive and the politics of exemptions. *Water* 8: 388–403. https://doi.org/10.3390/w8090388.

Bomans, E., Fransen, K., Gobin, A. et al. (2005). Addressing phosphorus related problems in farm practice. Final report to the European Commission. European Commission DG Environment. Project No P/OO/027, pp. 283.

Bouamra, F., Drouiche, N., Ahmed, D.S. et al. (2012). Treatment of water loaded with orthophosphate by electrocoagulation. *Procedia Engineering* 33: 155–162.

Boyd, J. (2000). *The New Face of the Clean Water Act: A Critical Review of the EPA's Proposed TMDL Rules*. Washington, DC: Resources for the Future pp. 37.

Boyd, J. (2003). Water pollution taxes: a good idea doomed to failure? Resources for the Future. May 2003. Discussion Paper 03–20. http://ageconsearch.umn.edu/bitstream/10611/1/dp030020.pdf (accessed 27 April 2019).

Bracmort, K.S., Engel, B.A., and Frankenberger, J.R. (2004). Evaluation of structural best management practices 20 years after installation: Black Creek watershed, Indiana. *Journal of Soil and Water Conservation* 59 (5): 191–196.

Brady, N.C. and Weil, R.R. (1996). *Elements of the Nature and Properties of Soils*, 2e. New Jersey: Prentice Hall 740 pp.

Brand, L.E., Pablo, J., Compton, A. et al. (2010). Cyanobacterial blooms and the occurrence of the neurotoxin beta-N-methylamino-L-alanine (BMAA) in South Florida Aquatic Food Webs. *Harmful Algae* 9 (6): 620–635.

Briand, J.F., Jacquet, S., Bernard, C. et al. (2003). Review article health hazards for terrestrial vertebrates from toxic cyanobacteria in surface water ecosystems. *Veterinary Research* 34: 361–377.

Brix, H., Arias, C.A., and Del Bubba, M. (2001). Media selection for sustainable phosphorus removal in subsurface flow constructed wetlands. *Water Science and Technology* 44 (11–12): 47–54.

Brockett, J. (2016). Putting the focus on phosphorus. Utility Week Online (29 February). https://utilityweek.co.uk/putting-the-focus-on-phosphorus/ (accessed 10 May 2015).

Brogowski, Z. and Renman, G. (2004). Characterisation of Opoka as a basis for its use in wastewater treatment. *Polish Journal of Environmenatl Studies* 13 (1): 15–20.

Brooks, A.S., Rozenwald, M.N., Geohring, L.D. et al. (2000). Phosphorus removal by wollastonite: a constructed wetland substrate. *Ecological Engineering* 15: 121–132.

Brooks, K., Relve, K., and Veinla, H. (2013). Estonian law on controlling emissions of nutrients in the Baltic Sea region. The research project conducted at the Faculty of Law, Stockholm University 2012-2013. https://www.su.se/polopoly_fs/1.173830.1396887222!/menu/standard/file/Estonian%20country%20study%20for%20publishing.pdf (accessed 27 April 2019).

Brown, D. (2017). End-of-tile phosphorous removal system project final report. Stone Environmental Vermont Project Number 14-084. Report prepared for Vermont USDA NRCS and Friends of Northern Lake Champlain. 73 pp.

Browning, M. Lawman, M. and Steele, J. (1996). Best management practices to reduce diffuse pollution from agriculture. RD Technical Report P40. Environment

Agency R & D Dissemination centre, Bristol, UK. https://www.gov.uk/government/uploads/system/uploads/attachment_data/file/290207/str-p40-e-e.pdf (accessed 27 April 2019).

Brownlie, W. (2014). Assessing the role of domestic phosphorus emissions in the human phosphorus footprint. Doctoral thesis. Heriot Watt University.

Buckwell, A. and Nadeau, E. (2016). *Nutrient Recovery and Reuse (NRR) in European Agriculture. A Review of the Issues, Opportunities, and Action*. Brussels: The Rural Investment Support for Europe (RISE) Foundation www.risefoundation.eu/images/files/2016/2016_RISE_NRR_Full_EN.pdf (accessed 27 April 2019).

Buda, A.R., Koopmans, G.F., Bryant, R.B. et al. (2012). Emerging technologies for removing nonpoint phosphorus from surface water and groundwater: introduction. *Journal of Environmental Quality* 41: 621–627.

Bugbee, P. (2015). Ag runoff needs attention; mandated buffers won't help. Times Writers Group. https://www.sctimes.com/story/opinion/2015/03/22/ag-runoff-needs-attention-mandated-buffers-help/25184729/ (accessed 27 April 2019).

Bullen, C. (2017). Phosphorus removal in the catchment. The BIG Phosphorous Conference and Exhibition, Manchester, 4–5 July 2017. https://www.aquaenviro.co.uk/wp-content/uploads/2017/09/Chris-Bullen-Siltbuster.pdf (accessed 27 April 2019).

Bunce, J.T., Ndam, E., Ofiteru, I.D. et al. (2018). A review of phosphorus removal technologies and their applicability to small-scale domestic wastewater treatment systems. *Frontiers in Environmental Science* 6 (8): 1–16. https://doi.org/10.3389/fenvs.2018.00008.

Burns, R.T. and Moody, L.B. (2002). Phosphorus recovery from animal manures using optimized struvite precipitation. *Proceedings of Coagulants and Flocculants: Global Market and Technical Opportunities for Water Treatment Chemicals*, Chicago, Illinois, 22–24 May 2002.

Butler, E., Hung, Y.-T., Yeh, R.Y. et al. (2011). Electrocoagulation in wastewater treatment. *Water* 3: 495–525. https://doi.org/10.3390/w3020495.

Canadian Council of the Ministers of the Environment (CCME) (2006). Examination of the potential funding mechanisms for municipal wastewater effluents (MWWE) projects in Canada. https://www.ccme.ca/files/Resources/municipal_wastewater_efflent/mwwe_funding_mechanisms_rpt_e.pdf (accessed 27 April 2019).

Carpenter, S.R. and Lathrop, R.C. (2008). Probabilistic estimate of a threshold for eutrophication. *Ecosystems* 11: 601–613.

Carstensten, J., Andersen, J.H., Gustafsson, B.G. et al. (2014). Deoxygenation of the Baltic Sea during the last century. *Proceedings of the National Academy of Sciences of the United States of America (PNAS)* 11 (15): 5628–5633. www.pnas.org/cgi/doi/10.1073/pnas.1323156111 (accessed 27 April 2019).

de Carvalho Aguiar, V.M., Neto, J.A.B., and Rangel, C.M. (2011). Eutrophication and hypoxia in four streams discharging in Guanabara Bay, Brazil, a case study. *Marine Pollution Bulletin* 62 (8): 1915–1919.

CEEP (1998). Conference summary, First International Conference on the recovery of phosphorus from sewage and animal wastes, Warwick University, UK. 6–7th May 1998. https://phosphorusplatform.eu/images/download/Warwick-1998-1st-P-Recovery-and-Recycling-conference-summary--papers.pdf (accessed 27 April 2019).

Charles, D. (2017). The Gulf of Mexico's dead zone is the biggest ever seen. https://www.npr.org/sections/thesalt/2017/08/03/541222717/the-gulf-of-mexicos-dead-zone-is-the-biggest-ever-seen (accessed 27 April 2019).

Chaturvedi, S.I. (2013). Electrocoagulation: a novel waste water treatment method. *International Journal of Modern Engineering Research (IJMER)* 3 (1): 93–100. http://www.ijmer.com/papers/Vol3_Issue1/AP3193100.pdf (accessed 27 April 2019).

Chazarenc, F., Kacem, M., Gérente, C. et al. (2008). 'Active' filters: a mini-review on the use of industrial by-products for upgrading phosphorus removal from treatment wetlands. Proceedings of the 11th International Conference on Wetland Systems for Water Pollution Control, Indore, India, 1–7 November 2008.

Cherry, K.A., Shepherd, M., Withers, P.J.A. et al. (2008). Assessing the effectiveness of actions to mitigate nutrient loss from agriculture: a review of methods. *The Science of Total Environment* 406 (1–2): 1–23.

Christian-Smith, J., Gleick, P.H., Cooley, H. et al. (2012). *A Twenty-First Century US Water Policy*. Oxford: Oxford University Press. 360 pp.

Coffman, L.S. (2000). Low-impact development design: a new paradigm for stormwater management mimicking and restoring the natural hydrologic regime; an alternative stormwater management technology. Maryland County, USA: Prince George's County Department of Environmental Resources and US EPA (Report EPA 841-B-00-003).

Coffmann, L.S., Goo, R. and Frederick, R.P.E. (1999). Low-impact development: an innovative alternative approach to stormwater management. 29th Annual Water Resources Planning and Management Conference, 6–9 June 1999, Tempe, AZ, USA. https://ascelibrary.org/doi/abs/10.1061/40430%281999%29118?src=recsys (accessed 27 April 2019).

Collins, R., Mcleod, M., Hedleyc, M. et al. (2007). Best management practices to mitigate faecal contamination by livestock of New Zealand waters. *New Zealand Journal of Agricultural Research* 50 (2): 267–278.

Conservation Law Foundation (2018). Lake Champlain Legal History. https://www.clf.org/legal-history/ (accessed 4 May 2018).

Contech Engineered Solutions (2018). The stormwater management stormfilter. http://www.contech-cpi.com/Products/Stormwater-Management/Treatment/Stormwater-Management-StormFilter.aspx (accessed 14 December 2018).

Convergent Water Technologies (2018). Convergent Water Technologies. www.convergentwater.com (accessed 27 April 2019).

Copeland, C., Maguire, S., and Mallett, W.J. (2016). *Legislative Options for Financing Water Infrastructure*. Washington: Congressional Research Service http://nationalaglawcenter.org/wp-content/uploads/assets/crs/R42467.pdf.

Cordell, D., Drangert, J.-O., and White, S. (2009). The story of phosphorus: global food security and food for thought. *Global Environmental Change* 19: 293–305.

Cordell, D., Rosemarin, A., Schröder, J.J. et al. (2011). Towards global phosphorus security: a systems framework for phosphorus recovery and reuse options. *Chemosphere* 84: 747–758.

Cornel, P. and Schaum, C. (2009). Phosphorus recovery from wastewater: needs, technologies and costs. *Water Science & Technology* 59 (6): 1069–1076.

Correll, D.L. (1998). The role of phosphorus in the eutrophication of receiving waters: a review. *Journal of Environmental Quality* 27: 261–266.

Cost869 (2018). Mitigation options for nutrient reduction in surface water and groundwaters. www.cost869.alterra.nl (accessed 27 April 2019).

Coxe, H.M. and Hedrich, M.F. (2007). *Manual of Best Management Practices for Maine Agriculture*. Augusta: Maine Department of Agriculture, Food & Rural Resources Division of Animal Health & Industry, pp. 107. Available at: http://www.maine.gov/dacf/php/nutrient_management/documents/BMP-Manual-Final-January-2007.pdf (accessed 27 April 2019).

Cucarella, C.V. (2009). Recycling filter substrates used for phosphorus removal from wastewaters as soil amendments. TRITA-LWR PhD thesis 1049. Royal Institute of Technology KTH Stockholm, TH, Sweden.

Cucarella, V. and Renman, G. (2009). Phosphorus sorption capacity of filter materials used for on-site wastewater treatment determined in batch experiments – a comparative study. *Journal of Environmental Quality* 38 (2): 381–392.

Cucarella, V., Zaleski, T., and Mazurek, R. (2007). Phosphorus sorption capacity of different types of opoka. *Annals of Warsaw University of Life Sciences – SGGW Land Reclamation* 38: 11–18.

Cullen, N., Baur, R., and Schauer, P. (2013). Three years of operation of North America's first nutrient recovery facility. *Water Science and Technology* 68 (4): 763–768. https://doi.org/10.2166/wst.2013.260.

Cuttle, S.P., Macleod, C.J.A., Chadwick, D.R. et al. (2007). An inventory of methods to control diffuse water pollution from agriculture: user manual. DEFRA project ES0203. London: DEFRA.

Daneshgar, S., Callegari, A., Capodaglio, A.G. et al. (2018). The potential phosphorus crisis: resource conservation and possible escape technologies: a review. *Resources* 7 (2): 37. https://doi.org/10.3390/resources7020037, https://www.mdpi.com/2079-9276/7/2/37/htm (accessed 27 April 2019).

Davis, A.P., Hunt, W.F., Traver, R.G. et al. (2009). Bioretention technology: overview of current practice and future needs. *Journal of Environmental Engineering* 135: 109–117.

De Feo, G., Antoniou, G., Fardin, F.H. et al. (2014). The historical development of sewers worldwide. *Sustainability* 6: 3936–3974. https://doi.org/10.3390/su6063936.

De Wit, J., van Keulen, H., van der Meer, H.G. et al. (1997). Animal manure: asset or liability? *World Animal Review* 88: 30–37.

DeBusk, K. and Wynn, T. (2011). Storm-water bioretention for runoff quality and quantity mitigation. *Journal of Environmental Engineering* 137 (9): 800–808.

Depurgan (2019). Pig manure treatment system. http://www.depurgan.com/ingles/index.php (accessed 27 April 2019).

Desmidt, E., Ghyselbrecht, K., Zhang, Y. et al. (2015). Global phosphorus scarcity and full-scale p-recovery techniques: a review. *Critical Reviews in Environmental Science and Technology* 45: 336–384. https://doi.org/10.1080/10643389.2013.866531.

Diers, J. (2013). Chautauqua lake TMDL for phosphorus. http://wri.cals.cornell.edu/sites/wri.cals.cornell.edu/files/shared/documents/Chautauqua_May_Diers.pdf (accessed 27 April 2019).

Dietz, M.E. and Clausen, J.C. (2005). A field evaluation of rain garden flow and pollutant treatment. *Water Air Soil Pollution* 167 (1–4): 123–138. https://doi.org/10.1007/s11270-005-8266-8.

van Dijk, K. (2018). ESPP EU nutrient research & development projects list. https://phosphorusplatform.eu/images/download/ESPP-EU-RD-nutrient-projects-list_v2018-09-14.pdf (accessed 27 April 2019).

Dodds, W.K., Bouska, W.W., Eitzmann, J.L. et al. (2009). Eutrophication of US freshwaters: analysis of potential economic damages. *Environmental Science & Technology* 43: 12–19.

Dodds, W.K., Jones, J.R., and Welch, E.B. (1998). Suggested classification of stream trophic state: distributions of temperate stream types by chlorophyll, total nitrogen and phosphorus. *Water Research* 32 (5): 1455–1462.

Dosskey, M.G. (2002). Setting priorities for research on pollution reduction functions of agricultural buff ers. *Environmental Management* 30: 641–650. https://doi.org/10.1007/s00267-002-2755-y.

Dou, Z., Ferguson, J.D., Fiorini, J. et al. (2003). Phosphorus feeding levels and critical control points on dairy farms. *Journal of Dairy Science* 86 (11): 3787–3795.

Douglas, G.B. (2002). Remediation material and remediation process for sediments. US Patent 6350383.

Drizo, A. (2010). Innovative technologies for phosphorus reduction from non point pollution sources. ASA, CSSA, SSSA 2010 International Meetings, Long Beach, CA, 31 October–3 November 2010.

Drizo, A. (2011a). Phosphorus and *E.Coli* reduction from silage leachate via innovative steel slag filtration. Conservation Innovation Grants Final Progress Report to the USDA, December, 2011. Grant Agreement Number Agreement Number: 69-3A75-9-121.

Drizo, A. (2011b). Phosphorous and suspended solids reduction from agricultural tile drainage via passive filtration systems. Final report submitted to the VT Environmental Protection Agency (EPA) Section 319 Nonpoint Source Grants Program, September, 2011.

Drizo, A. (2011c). Phosphorus and *E. Coli* reduction from silage leachate via innovative passive filtration systems. Final report submitted to VT NRCS Conservation Innovation Grants Program and VAAFM, September 2011.

Drizo, A. (2012). Innovative phosphorus removal technologies. *Australian Journal of Clean Technologies* http://www.azocleantech.com/article.aspx?ArticleID=226 (accessed 27 April 2019).

Drizo, A. (2013). Phosphorus removal, resource recovery and reuse from onsite septic systems. Presentation distributed at the Scientific and Technical Advisory Committee (STAC) workshop 'In My Backyard: An Innovative Look at the Advances of Onsite Decentralized Wastewater Treatment', Washington DC, 17–18 December 2013.

Drizo, A. (2017). Solving eutrophication: past, present, and future. Invited Guest lecture at Concordia University Loyola Sustainability Research Centre. Montreal, Canada, 10 November 2017. http://www.concordia.ca/cuevents/artsci/loyola-college/2017/11/10/solving-eutrophication--past--present--and-future-by-dr--aleksan.html (accessed 27 April 2019).

Drizo, A., Cummings, J., Weber, D. et al. (2008b). New evidence for rejuvenation of phosphorus retention capacity in EAF steel slag. *Environmental Science and Technology* 42: 6191–6197.

Drizo, A., Forget, C., Chapuis, R.P. et al. (2000). How realistic are the linear Langmuir predictions of phosphate retention by adsorbing materials? Proceedings of the 1st World Congress of the International Water Association held in Paris, 3–7 July 2000. London: IWA publishing.

Drizo, A., Forget, C., Chapuis, R.P. et al. (2002). Phosphorus removal by EAF steel slag – a parameter for the estimation of the longevity of constructed wetland systems. *Environmental Science and Technology* 36: 4642–4648.

Drizo, A., Forget, C., Chapuis, R.P. et al. (2006). Phosphorus removal by electric arc furnace (EAF) steel slag and serpentinite. *Water Research* 40 (8): 1547–1554.

Drizo, A., Frost, A.C., Smith, K.A. et al. (1997). The use of constructed wetlands in phosphate and ammonium removal from wastewater. *Water Science and Technology* 35 (5): 95–102.

Drizo, A., Frost, A.C., Smith, K.A. et al. (1999). Physico-chemical screening of phosphate-removing substrates for use in constructed wetland systems. *Water Research* 33 (17): 3595–3602.

Drizo, A., Gunes, K., and Picard, H. (2016). PhosphoReduc technology for phosphorus removal and harvesting from onsite septic systems – a novel strategy for eutrophication mitigation and control. 10th European Wastewater Management Conference, 11–12 October 2016, Manchester, UK.

Drizo, A. and Picard, H. (2014). Systems and methods for removing phosphorus from wastewaters. US Serial number 12/807, 177. S/N: 12/807,177; US20012/0048806, 1 March 2012. US 20120048806 A1, Published 13 May 2014.

Drizo, A. and Picard, H. (2015). Phosphorus harvesting, recycling and re-use via PhosphoReduc technology. Second European Sustainable Phosphorus Conference (ESPC2), 5–6 March 2015, Berlin, Germany.

Drizo, A., Seitz, E., Twohig, E. et al. (2008a). The role of vegetation in phosphorus removal by cold climate constructed wetland: the effects of aeration and growing season. In: *Wastewater Treatment,* Plant Dynamics and Management in Constructed and Natural Wetlands (ed. J. Vymazal), 237–251. Dordrecht: Springer Publishing.

Duda, A.M. and Finan, D.S. (1983). Influence of livestock on nonpoint source nutrient levels of streams. *Transactions of the American Society of Agricultural Engineers (ASAE)* 26: 1710–1716.

Dudley, B. and May, L. (2007). *Estimating the Phosphorus Load to Water Bodies From Septic Tanks.* Bailrigg: Centre for Ecology and Hydrology, 45pp. (CEH Project Number: C03273, C01352).

Dura, A. (2013). Electrocoagulation for water treatment: the removal of pollutants using aluminium alloys, stainless steels and iron anodes. PhD thesis. National University of Ireland Maynooth.

Ecofiltration Nordic (2018). About ecofilration nordic. http://www.ecofiltration.se/en/about-ecofiltration (accessed 27 April 2019).

Edwards, D.R. and Daniel, T.C. (1992). Environmental impacts of on-farm poultry waste-disposal – a review. *Bioresource Technology* 41: 9–33.

Eggers, E., Dirkzwager, A.H., and van der Honing, H. (1991). Full-scale experiences with phosphate crystallization in a crystalactor. *Water Science and Technology* 23 (4–6): 819–824. https://doi.org/10.2166/wst.1991.0533.

Egle, L., Rechberger, H., Krampe, J. et al. (2016). Phosphorus recovery from municipal wastewater: an integrated comparative technological, environmental and economic assessment of P recovery technologies. *Science of the Total Environment* 571: 522–542. http://dx.doi.org/10.1016/j.scitotenv.2016.07.019.

*Encyclopedia Britannica* (2018a). Oxisol. https://www.britannica.com/search?query=oxisols (accessed 27 April 2019).

*Encyclopedia Britannica* (2018b). Ultisol. https://www.britannica.com/search?query=Ultisols (accessed 27 April 2019).

Ends waste and bioenergy (2015). Swedish 'DeBugger' pilot plant revealed. Published June 24th, 2015. https://www.endswasteandbioenergy.com/article/1353167/swedish-debugger-pilot-plant-revealed (accessed 27 April 2019).

Environment and Climate Change Canada (2017). Evaluation of the Lake Winnipeg Basin initiative. Final report. Audit and Evaluation Branch. June 2017. https://www.canada.ca/content/dam/eccc/documents/pdf/evaluation-lake-winnipeg-basin/Evaluation_of%20_the_Lake_Winnipeg_Basin_Initiative.pdf (accessed 10 April 2019).

ETV Canada (2018). Process overview. http://etvcanada.ca/begin-your-etv-screening-application-here/ (accessed 17 December 2018).

EurEau (2019). The governance of water services in Europe. http://www.eureau.org/resources/publications/150-report-on-the-governance-of-water-services-in-europe/file (accessed 9 January 2019).

European Commission (2013). Research and innovation. Call: water innovation: boosting its value for Europe. https://ec.europa.eu/research/participants/portal/desktop/en/opportunities/h2020/calls/h2020-water-2014-2015.html (accessed 27 April 2019).

European Commission (2014). 20 critical raw materials – major challenge for EU industry. Press release 26 May 2014. http://europa.eu/rapid/press-release_IP-14-599_en.htm (accessed 27 April 2019).

European Commission (2015a). A water blueprint – taking stock, moving forward. The blueprint to safeguard Europe's water resources – communication from the commission COM (2012)673. http://eur-lex.europa.eu/legal-content/EN/TXT/PDF/?uri=CELEX:52012DC0673&from=EN (accessed 27 April 2019).

European Commission (2015b). *Best Practices in Improving the Sustainability of Agriculture*. Europa Joint Research Center https://ec.europa.eu/jrc/en/event/conference/best-practices-improving-sustainability-agriculture (accessed 27 April 2019).

European Commission (2015c). CAP explained direct payments for farmers 2015–2020. https://ec.europa.eu/agriculture/sites/agriculture/files/direct-support/direct-payments/docs/direct-payments-schemes_en.pdf (accessed 27 April 2019).

European Commission (2015d). Closing the loop: Commission adopts ambitious new circular economy package to boost competitiveness, create jobs and generate sustainable growth. Press release, 2 December, 2015. http://europa.eu/rapid/press-release_IP-15-6203_en.htm (accessed 20 January 2019).

European Commission (2015e). The EU pilot programme on environmental technology verification (ETV). http://ec.europa.eu/environment/gpp/pdf/06_10_2015/EU_ETV_and_GPP.pdf (accessed 27 April 2019).

European Commission (2016a). Environment. http://ec.europa.eu/environment/archives/etv/links.htm (accessed 27 April 2019).

European Commission (2016b). New regulation to boost the use of organic and waste-based fertilisers. https://ec.europa.eu/growth/content/new-regulation-boost-use-organic-and-waste-based-fertilisers-0_en (accessed 27 April 2019).

European Commission (2016c). Urban Waste Water Directive. http://ec.europa.eu/environment/water/water-urbanwaste/legislation/directive_en.htm (accessed 15 April 2017).

European Commission (2017). Agriculture and Water. https://ec.europa.eu/agriculture/envir/water_en (accessed 27 April 2019).

European Commission (2018a). *Best environmental management practice for the agriculture sector – crop and animal production*. Brussels: European Commission. Final draft at: http://susproc.jrc.ec.europa.eu/activities/emas/documents/AgricultureBEMP.pdf (accessed 27 April 2019).

European Commission (2018b). Eco_innovation action plan. https://ec.europa.eu/environment/ecoap/about-eco-innovation_en (accessed 27 April 2019).

European Commission (2018c). Circular economy: agreement on Commission proposal to boost the use of organic and waste-based fertilisers. Press release 12 December, 2018, Brussels. http://europa.eu/rapid/press-release_IP-18-6161_en.htm (accessed 27 April 2019).

European Commission (2018d). Circular economy: new rules will make EU the global front-runner in waste management and recycling. Press release 22 May 2018, Brussels. http://europa.eu/rapid/press-release_IP-18-3846_en.htm (accessed 27 April 2019).

European Commission (2018e). EU ETV pilot programme: three years of supporting innovation. https://ec.europa.eu/environment/ecoap/etv/news/eu-etv-pilot-programme-three-years-supporting-innovation_en (accessed 27 April 2019).

European Commission (2018f). Horizon 2020. https://ec.europa.eu/programmes/horizon2020/en/search/site/Water%2520 (accessed 27 April 2019).

European Commission (2018g). TOPIC: EU-India water co-operation. http://ec.europa.eu/research/participants/portal/desktop/en/opportunities/h2020/topics/sc5-12-2018.html (accessed 27 April 2019).

European Court of Auditors (2015). Special report. EU-funding of urban waste water treatment plants in the Danube river basin: further efforts needed in helping Member States to achieve EU waste water policy objectives. https://www.eca.europa.eu/Lists/ECADocuments/SR15_02/SR_DANUBE_RIVER_EN.pdf (accessed 27 April 2019).

European Environmental Agency (EAA) (1995). *Europe's Environment: The Dobris Assessment*. Copenhagen, Denmark: European Environment Agency.

European Environmental Agency (EEA) (2015). Urban waste water treatment map. https://www.eea.europa.eu/themes/water/water-pollution/uwwtd/interactive-maps/urban-waste-water-treatment-maps (accessed 27 April 2019).

European Parliament and Council (2006). Directive 2006/118/EC of the European Parliament and of the Council of 12 December 2006 on the protection of groundwater against pollution and deterioration (Daughter of 2000/60/EC). https://eur-lex.europa.eu/legal-content/EN/TXT/?uri=CELEX%3A32006L0118 (accessed 27 April 2019).

Eveborn, D. (2013). Sustainable phosphorus removal in onsite wastewater treatment. PhD thesis. Royal Institute of Technology KTH, Stockholm, Sweden. http://kth.diva-portal.org/smash/get/diva2:616601/FULLTEXT01.pdf (accessed 27 April 2019).

Facts and Details (2014). Water pollution in China. http://factsanddetails.com/china/cat10/sub66/item391.html (accessed 27 April 2019).

Falk, M.W., Reardon, D.J., Neethling, J.B. et al. (2013). Striking the balance between nutrient removal, greenhouse gas emissions, receiving water quality, and costs. *Water Environment Research* 85 (12): 2307–2316.

Fane, A.G. and Fane, S.A. (2005). The role of membrane technology in sustainable decentralized wastewater systems. *Water Science & Technology* 51 (10): 317–325.

FAO (2003). Summary analysis of codes, guidelines, and standards related to good agricultural practices: background paper for the FAO expert consultation on a good agricultural practice approach. http://www.fao.org/docrep/pdf/010/ag853e/ag853e00.pdf (accesed 27 April 2019).

FAO (2013). Guidelines to control water pollution from agriculture in China: decoupling water pollution from agricultural production. FAO Water Report 40. http://www.fao.org/docrep/019/i3536e/i3536e.pdf (accesed 27 April 2019).

FAO (2015a). Livestock Policy Brief 02: Pollution from industrialized livestock production. Report by the Livestock Information, Sector Analysis and Policy Branch Animal Production and Health Division. http://www.fao.org/3/a-a0261e.pdf (accesed 27 April 2019).

FAO (2015b). World agriculture: towards 2015/2030. An FAO perspective on livestock production. FAO Corporate Document Repository. Economic and Social Development Department. http://www.fao.org/3/y4252e/y4252e00.htm#TopOfPage (accesed 27 April 2019).

FAO (2015c). Are grasslands under threat? Brief analysis of FAO statistical data on pasture and fodder crops. http://www.fao.org/uploads/media/grass_stats_1.pdf (accesed 27 April 2019).

Fistarol, G.O., Coutinho, F.H., Moreira, A.P.B. et al. (2015). Environmental and sanitary conditions of Guanabara Bay, Rio de Janeiro. *Frontiers in Microbiology* 6: 1232–1249. https://doi.org/10.3389/fmicb.2015.01232.

Flesher, J. (2017). Farm runoff and the worsening algae plague. Phys.Org. https://phys.org/news/2017-11-farm-runoff-worsening-algae-plague.html (accessed 27 April 2019).

Fletcher, T.D., Shuster, W., Hunt, W.F. et al. (2015). SUDS, LID, BMPs, WSUD and more – the evolution and application of terminology surrounding urban drainage. *Urban Water Journal* 12 (7): 525–542. https://doi.org/10.1080/1573062X.2014.916314.

Foged, H., Flotats, X., Bonmati Blasi, A. et al. (2011). Inventory of manure processing activities in Europe. Technical Report No. I concerning 'Manure Processing Activities in Europe' to the European Commission, Directorate-General Environment. 138 pp. http://agro-technology-atlas.eu/docs/21010_technical_report_I_inventory.pdf (accessed 27 April 2019).

Forbes, E.G.A., Foy, R.H., Mulholland, M.V. et al. (2011). Performance of a constructed wetland for treating farm-yard dirty water. *Water Science & Technology* 64 (1): 22–28.

Forget, C. (2001). Élimination du phosphore dissous des effluents piscicoles à` l'aide de matériaux granulaires réactifs. MSc thesis. Ecole Polytechnique of Montreal, Canada.

Foundation for Water Research (2006). Eutrophication of freshwaters. Report FR/R0002, pp. 69. http://www.fwr.org/eutrophi.pdf (accessed 27 April 2019).

Föyn, E. (1964). Removal of sewage nutrients by electrolytic treatment. *Verhandlungen des Internationalen Verein Limnologie* 15: 569–579.

Francis, G. (1878). Poisonous Australian lake. *Nature* 18: 11–12.

Franzetti, S.M. (2016). *Background and History of Stormwater Regulations*. New York: Franzetti Law Firm PC.

Fraunhofer-Gesellschaft (2018). Innovative process for environmentally friendly manure treatment comes onto the market. https://phys.org/news/2018-05-environmentally-friendly-manure-treatment.html (accessed 27 April 2019).

Friedrich, J., Janssen, F., Aleyink, D. et al. (2014). Investigating hypoxia in aquatic environments: diverse approaches to addressing a complex phenomenon. *Biogeosciences* 11: 1215–1259. https://doi.org/10.5194/bg-11-1215-2014.

Fuji Clean (2018). Phosphorus removal device for residential unit. https://www.fujiclean.co.jp/fujiclean/english/technology/index.html (accessed 11 November 2018).

Gaddis, E.J.B., Voinov, A., Seppelt, R. et al. (2014). Spatial optimization of best management practices to attain water quality targets. *Water Resources Management* 28: 1485–1499.

Gasiorowski-Denis, E. (2016). New ISO tool to help cutting-edge green technologies reach markets. https://www.iso.org/news/2016/11/Ref2144.html (accessed 27 April 2019).

Georgakakos, C.B., Morris, C.K., and Walter, M.T. (2018). Challenges and opportunities with on-farm research: total and soluble reactive stream phosphorus before and after implementation of a cattle-exclusion, riparian buffer. *Frontiers in Environmental Science* 05 July 2018. https://doi.org/10.3389/fenvs.2018.00071.

Ghosh, T.K. and Mondal, D. (2012). Eutrophication: causative factors and remedial measures. *Journal of Today's Biological Sciences: Research & Review* 1 (1): 153–178.

Gill, L.W. (2011). The development of a code of practice for single house on-site wastewater treatment in Ireland. *Water Science and Technology* 64 (3): 677–683.

Gilliom, R.J. and Patmont, C.R. (1983). Lake phosphorus loading from septic systems by seasonally perched groundwater. *Journal of the Water Pollution Control Federation* 55 (10): 1297–1305.

van Ginkel, C.E. (2011). Eutrophication: present reality and future challenges for South Africa. *Water SA* 37 (5): 693–701. http://dx.doi.org/10.4314/wsa.v37i5.6.

Gold, A. and Sims, J. (2006). Phosphorus from septic tanks. Scope Newsletter 63 Special issue: fate of phosphorus in septic tanks. https://phosphorusplatform.eu/images/download/ScopeNewsletter63.pdf (accessed 27 April 2019).

Goswami, N. (2018). Farmers say they're being unfairly targeted over pollution in Lake Carmi. http://www.wcax.com/content/news/Farmers-say-theyre-not-to-blame-for-Lake-Carmi-pollution-479560383.html (accessed 27 April 2019).

Griffiths, M. (2002). *The European Water Framework Directive: An Approach to Integrated River Basin Management*. Hennef: European Water Management Online http://www.ewa-online.eu/tl_files/_media/content/documents_pdf/Publications/E-WAter/documents/85_2002_05.pdf (accessed 27 April 2019).

Gulliver, J.S. (2014). Assessing and improving pollution prevention by Swales. Minnesota Department of Transportation. Research Project Final Report 2014-30. August 2014. https://conservancy.umn.edu/handle/11299/168022 (accessed 27 April 2019).

Gunes, K., Tuncsiper, B., Ayaza, S. et al. (2012). Free water surface constructed wetland system for treatment of high strength domestic wastewater: a case study for the Mediterranean. *Ecological Engineering* 44: 278–284.

Haan, M.M., Russell, J.R., Powers, W.J. et al. (2006). Grazing management effects on sediment and phosphorus in surface runoff. *Rangeland Ecology Management* 59: 607–615. https://doi.org/10.2111/05-152R2.1.

Habibiandehkordi, R., Lobb, D.A., Owens, P.N. et al. (2018). Effectiveness of vegetated buffer strips in controlling legacy phosphorus exports from agricultural land. *Journal of Environmental Quality* 48 (2): 314–321. https://doi.org/10.2134/jeq2018.04.0129.

Han, D., Currell, M.J., and Cao, G. (2016). Deep challenges for China's war on water pollution. *Environmental Pollution* 218: 1222–1233.

Hart, M.R., Quin, B.F., and Nguyen, M.L. (2004). Phosphorus runoff from agricultural land and direct fertilizer effects: a review. *Journal of Environmental Quality* 33: 1954–1972.

Hautier, Y., Niklaus, P.A., and Hector, A. (2009). Competition for light causes plant biodiversity loss after eutrophication. *Science* 324 (5927): 636–638. https://doi.org/10.1126/science.1169640.

Havukainen, J., Nguyen, M.T., Hermann, L. et al. (2016). Potential of phosphorus recovery from sewage sludge and manure ash by thermochemical treatment. *Waste Management* 49: 221–229. https://doi.org/10.1016/j.wasman.2016.01.020.

HDR (2007). Spokane county onsite sewage disposal systems. Phosphorus Loading Estimate Technical Memorandum. 27 June 2007. https://www.spokanecounty.org/DocumentCenter/View/2001/B-Septic---Phosphorus-Study---Final-PDF (accessed 27 April 2019).

Heathwaite, A.L. and Dils, R.M. (2000). Characterising phosphorus loss in surface and subsurface hydrologic pathways. *The Science of the Total Environment* 252: 523–538.

Hebert, C. (2013). Une spécialiste met en doute l'efficacité des bandes riveraines. L'Avenir, 9 October 2013. http://www.laveniretdesrivieres.com/Actualites/2013-10-09/article-3422198/Une-specialiste-met-en-doute-lefficacite-des-bandes-riveraines/1 (accessed 27 April 2019).

Heckenmüller, M., Narita, D., and Klepper, G. (2014). Global availability of phosphorus and its implications for global food supply: an economic overview. Kiel Working Paper No. 1897. January 2014. https://pdfs.semanticscholar.org/4633/84ce6af3320edde7f31cb6cd8637117c1b52.pdf (accessed 27 April 2019).

Heinen, M., Noij, I.G.A.M., Heesmans, H.I.M. et al. (2012). A novel method to determine buffer strip effectiveness on deep soils. *Journal of Environmental Quality* 41: 334–347. https://doi.org/10.2134/jeq2010.0452.

Hilliard, C., Scott, N., Lessa, A. et al. (2002). Agricultural best management practices for the Canadian prairies. Agriculture and Agri Food Canada File No.: 6672-1-12-1-18, March 31, 2002, pp. 72. http://www.uach.cl/externos/epicforce/pdf/guias%20y%20manuales/varios/hilliard_C_et_al_2002.pdf (accessed 27 April 2019).

Hirschman, D.J., Seipp, B., and Schueler, T. (2017). Performance enhancing devices for stormwater best management practices. Final Report. https://mostcenter.org/sites/default/files/performance_enhancing_devices_stormwater_bmps.pdf (accessed 27 April 2019).

Hong, K.H., Chang, D., Bae, H.S. et al. (2013). Electrolytic removal of phosphorus in wastewater with noble electrode under the conditions of low current and constant voltage. *International Journal of Electrochemical Science* 8: 8557–8571.

Horan, N.J. (1990). *Biological Wastewater Treatment Systems: Theory and Operation*. New York: Wiley.

Howarth, R.W. and Marino, R. (2006). Nitrogen as the limiting nutrient for eutrophication in coastal marine ecosystems: evolving views over three decades. *Limnology and Oceanography* 51 (1): 364–376.

Hsieh, C. and Yang, W. (2007). Optimal nonpoint source pollution control strategies for a reservoir watershed in Taiwan. *Journal of Environmental Management* 85: 908–917.

Hsieh, C.H., Davis, A.P., and Needelman, B.A. (2007). Bioretention column studies of phosphorus removal from urban stormwater runoff. *Water Environment Research* 79 (2): 177–184. PMID:17370843. doi:https://doi.org/10.2175/106143006X111745.

Hunt, P.G. and Poach, M.E. (2001). State of the art for animal wastewater treatment in constructed wetlands. *Water Science and Technology* 44 (11–12): 19–25.

Hunt, W.F., Jarrett, A.R., Smith, J.T. et al. (2006). Evaluating bioretention hydrology and nutrient removal at three field sites in North Carolina. *Journal of Irrigation and Drainage Engineering* 132 (6): 600–608. https://doi.org/10.1061/(ASCE)0733-9437(2006)132:6(600).

Hydro International (2009). Commercialization of upflow by hydro international. http://unix.eng.ua.edu/~rpitt/Presentations/Regional_Conferences/UpFlo_Stormwater_2009_WSUD.pdf (accessed 27 April 2019).

Hydro International (2018). Up-flo filter. https://www.hydro-int.com/en/products/flo-filter (accessed 27 April 2019).

Hylander, L.D., Kietlinska, A., Renman, G. et al. (2006). Phosphorus retention in filter materials for wastewater treatment and its subsequent suitability for plant production. *Bioresource Technology* 97: 914–921.

Hylander, L.D. and Siman, G. (2001). Plant-availability of phosphorus sorbed to potential wastewater treatment materials. *Biology and Fertility of Soils* 34: 28–42.

IC-IMPACTS (2018). India-Canada centre for innovative multidisciplinary partnerships to accelerate community transformation and sustainability. https://ic-impacts.com/about/ (accessed 27 April 2019).

Imbrium Systems (2007). Imbrium systems introduces SorbTIVE engineered media for stormwater pollutants. Pollution Online. https://www.pollutiononline.com/doc/imbrium-systems-introduces-sorbtive-engineere-0001 (accessed 27 April 2019).

Imbrium Systems (2010). Imbrium systems and Rinker materials team up on new filtration and phosphorus removal products to protect Massachusetts Watersheds. Pollution Online. https://www.pollutiononline.com/doc/imbrium-systems-and-rinker-materials-team-0001 (accessed 27 April 2019).

Indiana University Library (2015). Useful septic systems statistics. http://webapp1.dlib.indiana.edu/virtual_disk_library/index.cgi/5573230/FID669/Public%20education/Septic%20System%20facts.pdf (accessed 27 April 2019).

Informed Infrastructure (2014). Contech engineered solutions acquires assets of filterra bioretention systems. https://informedinfrastructure.com/10481/contech-engineered-solutions-acquires-assets-of-filterra-bioretention-systems (accessed 27 April 2019).

Inglezakis, V.J., Zorpas, A.A., Karagiannidis, A. et al. (2014). European Union legislation on sewage sludge management. *Fresenius Environmental Bulletin* 23 (2A): 635–639. ISSN: 10184619. CODEN: FENBE.

Innovationpolicyplatform (2018). STI Outlook Country profile 2016. https://www.innovationpolicyplatform.org/content/brazil (accessed 27 April 2019).

Interreg (2019). Phos4You – we deliver phosphorus 'made in Europe'. Interreg North-West Europe. http://www.nweurope.eu/projects/project-search/phos4you-phosphorus-recovery-from-waste-water-for-your-life (accessed 27 April 2019).

Jacobson, P.C., Hansen, G.J.A., Bethke, B.J. et al. (2017). Disentangling the effects of a century of eutrophication and climate warming on freshwater lake fish assemblages. *PLoS ONE* 12 (8): e0182667. https://doi.org/10.1371/journal.pone.0182667.

Jakobsson, C., Sommer, E.B., De Clercq, P. et al. (2002). The policy implementation of nutrient management legislation and effects in some European countries. http://siteresources.worldbank.org/INTAPCFORUM/Resources/lectureone7.doc (accessed 27 April 2019).

Jarvie, H.P., Sharpley, A.N., Withers, P.J.A. et al. (2013). Phosphorus mitigation to control river eutrophication: murky waters, inconvenient truths, and "postnormal" science. *Journal of Environmental Quality* 42: 295–304.

Jenssen, P.D., Krogstad, T., Paruch, A.M. et al. (2010). Filter bed systems treating domestic wastewater in the Nordic countries –performance and reuse of filter media. *Ecological Engineering* 36: 1651–1659.

Jiang, F., Beck, M.B., Cummings, R.G. et al. (2004). Estimation of costs of phosphorus removal in wastewater treatment facilities: construction de novo. Water Policy working paper #2004-010. http://www.h2opolicycenter.org/researchpapers/DeNovo.pdf (accessed 27 April 2019).

Jiang, F., Beck, M.B., Cummings, R.G. et al. (2005). Estimation of costs of phosphorus removal in wastewater treatment facilities: adaptation of existing facilities. Water Policy working paper #2005-011. https://www.issuelab.org/resources/4705/4705.pdf (accessed 27 April 2019)

Johansson, L. (1997). The use of LECA (light expanded clay aggregates) for the removal of phosphorus from wastewater. *Water Science and Technology* 35 (5): 87–93.

Johansson, L. (1999a). Industrial by-products and natural substrata as phosphorus sorbents. *Environmental Technology* 20: 309–316.

Johansson, L. (1999b). Blast furnace slag as phosphorus sorbents column studies. *The Science of the Total Environment* 229: 89–97.

Johansson, L. and Hylander, L. (1998). Phosphorus removal from waste water by filter media: retention and estimated plant availability of sorbed phosphorus. *Zeszyty Problemowe Postępów Nauk Rolniczych [Journal of Agricultral Science Problems]* 458: 397–409.

Johansson-Westholm, L. (2006). Substrates for phosphorus removal-potential benefits for on-site wastewater treatment? *Water Research* 40 (1): 23–36.

Johansson-Westholm, L. (2010). The use of blast furnace slag for removal of phosphorus from wastewater in Sweden—a review. *Water* 2010 (2): 826–837. https://doi.org/10.3390/w2040826.

Johansson-Westholm, L., Renman, G., and Drizo, A. (2010). The use of blast furnace and electric arc furnace steel slag in water pollution control. Conference Proceedings of the 6th European Slag Conference, Madrid, Spain, 2–4 October 2010.

Joko, I. (1985). Phosphorus removal from wastewater by the crystallization method. *Water Science and Technology* 17 (2–3): 121–132.

Kadlec, R. (2016). Large constructed wetlands for phosphorus control: a review. *Water* 8 (6): 243. https://doi.org/10.3390/w8060243.

Kaufmann, T. (2015). Sustainable livestock production: low emission farm – the innovative combination of nutrient, emission and waste management with special emphasis on Chinese pig production. *Animal Nutrition* 1: 104–112.

Kelessidis, A. and Stasinakis, A.S. (2012). Comparative study of the methods used for treatment and final disposal of sewage sludge in European countries. *Waste Management* 32 (6): 1186–1195. https://doi.org/10.1016/j.wasman.2012.01.012.

Kemp, R. (2001). Implementation of the Urban Wastewater treatment Directive (91/271/EEC) in Germany, the Netherlands, Spain, England and Wales. The tangible results. *Environmental Policy and Governance* 11 (5): 250–264. https://doi.org/10.1002/eet.272.

Khemka, R. (2016). From policy to practice to principles of water governance. *Economic & Political Weekly, EPW* 52: 27–31.

Khodadadi, M., Hosseinnejad, A., Rafati, L. et al. (2017). Removal of phosphate from aqueous solutions by iron nano-magnetic particle coated with powder activated carbon. *Journal of Health Science and Technology* 1 (1): 17–22.

King, K., Williams, M.R., Macrae, M.L. et al. (2014). Phosphorus transport in agricultural subsurface drainage: a review. *Journal of Environmental Quality* 44 (2): 467–485.

Kinley, R.D., Gordon, R.J., Stratton, G.W. et al. (2007). Phosphorus losses through agricultural tile drainage in Nova Scotia, Canada. *Journal of Environmental Quality* 36: 469–477.

Kinsley, C. and Joy, D. (2006). Field validation on an onsite wastewater risk assessment model final report to the Canada mortgage and housing corporation – January 2006. http://publications.gc.ca/collections/collection_2011/schl-cmhc/nh18-1/NH18-1-311-2005-eng.pdf (accessed 27 April 2019).

Kleinman, P.J.A., Sharpley, A.N., McDowell, R.W. et al. (2011). Managing agricultural phosphorus for water quality protection: principles for progress. *Plant and Soil* 349 (1–2): 169–182. https://doi.org/10.1007/s11104-011-0832-9.

Kleinman, P.J.A., Sharpley, A.N., Withers, P.J.A. et al. (2015). Implementing agricultural phosphorus science and management to combat eutrophication. *Ambio* 44 (Suppl. 2): S297–S310. https://doi.org/10.1007/s13280-015-0631-2.

Klimeski, A., Chardon, W.J., and Turtola, E. (2012). Potential and limitations of phosphate retention media in water protection: a process-based review of laboratory and field-scale tests. *Agricultural and Food Science* 21: 206–223.

Knight, R.L., Payne, V.W.E., Borer, R.E. et al. (2000). Constructed wetlands for live-stock wastewater management. *Ecological Engineering* 15: 41–55.

Koch, D., Lefler, M., and Britton, A. (2015). Phosphorus Recovery at the Stickney Water Reclamation Plant. https://www.mi-wea.org/docs/Koch_Lefler_Britton-Phosphorus_Recovery.pdf (accessed 27 April 2019).

Kostura, B., Kulveitova, H., and Lesko, J. (2005). Blast furnace slags as sorbents of phosphate from water solutions. *Water Research* 39 (9): 1795–1802. https://doi.org/10.1016/j.watres.2005.03.010.

Kraemer, R.A., Choudhury, K., and Kampa, E. (2001). Protecting water resources: pollution prevention. Proceedings of the International Conference on Freshwater, Bonn, Germany, 3–7 December 2001.

Kumar, M.D. and Ballabh, V. (2000). Water management problems and challenges in India. An analytical review. Institute of Rural Management Anand, India, Working Paper 140.

Kundu, S., Vassanda Coumar, M., Rajendiran, S. et al. (2015). Phosphates from detergents and eutrophication of surface water ecosystem in India. *Current Science* 108 (7): 1320–1325.

Lake Champlain Basin Program (LCBP) (2017). LCBP requests for proposals. http://www.lcbp.org/about-us/grants-rfps/request-for-proposals-rfps (accessed 27 April 2019).

Lam, Q.D., Schmalz, B., and Fohrer, N. (2011). The impact of agricultural Best Management Practices on water quality in a North German lowland catchment. *Environmental Monitoring Assessment* 183 (1-4): 351–379. https://doi.org/10.1007/s10661-011-1926-9.

Langmuir, I. (1918). The adsorption of gases on plane surfaces of glass, mica and platinum. *Journal of American Chemical Society* 40 (9): 1361–1403.

Le, C., Zha, Y., Li, Y. et al. (2010). Eutrophication of lake waters in China: cost, causes, and control. *Environmental Management* 45 (4): 662–668.

Lee, M., Drizo, A., Rizzo, D. et al. (2010). Evaluating the efficiency and temporal variation of pilot-scale hybrid and integrated constructed wetlands for treating high BOD and Phosphorus concentrated dairy effluent. *Water Research* 44 (14): 4077–4086.

Lee, S. et al. (2004). The transformation of the Shanghai Water sector in the reform era – social actors and institutional change. PhD dissertation. SOAS, University of London, March 2004.

Li, F.M., Liang, Z., Zhao, Z.W. et al. (2015). Toxicity of nano-TiO2 on algae and the site of reactive oxygen species production. *Aquatic Toxicology* 158: 1–13.

Lietman, P.L., Hall, D.W., Langland, M.J. et al. (1996). Evaluation of agricultural best management practices in the Conestoga River Headwaters, Pennsylvania. U.S. Geological Survey Water-Resources Investigations Report 93-411. http://pubs.usgs.gov/wri/1993/4119/report.pdf (accessed 27 April 2019).

Living Water Exchange (2018). EU database of best practices. http://archive.iwlearn.net/nutrient-bestpractices.iwlearn.org/nutrient-bestpractices.iwlearn.org/nutrient-bestpractices.iwlearn.org/index.html (accessed 12 September 2018).

Lockhart, J.A. (1997). Environmental tax policy in the United States: alternatives to the polluter pays principle. *Asia-Pacific Journal of Accounting* 4 (2): 219–239.

Lofrano, G. and Brown, J. (2010). Wastewater management through the ages: a history of mankind. *Science of the Total Environment* 408: 5254–5264.

Logan, T.J. (1990). Agricultural best management practices and groundwater quality. *Journal of Soil and Water Conservation* 45: 201–206.

Logan, T.J. (1993). Agricultural best management practices for water pollution control: current issues. *Agriculture, Ecosystems and Environment* 46: 223–231.

Loganathan, P., Vigneswaran, S., Kandasamy, J. et al. (2014). Removal and recovery of phosphate from water using sorption. *Critical Reviews in Environmental Science and Technology* 44 (8): 847–907. https://doi.org/10.1080/10643389.2012.741311.

Lombardo Associates Inc. (2018). Lombardo Associates website. www.lombardoassociates.com (accessed 12 September 2018).

Lu, H., Wang, J., Stoller, M. et al. (2016). An overview of nanomaterials for water and wastewater treatment. *Advances in Materials Science and Engineering* 2016: 1–10. http://dx.doi.org/10.1155/2016/4964828.

Lu, H., Wang, J., Stoller, M. et al. (2017). Crystallization techniques in wastewater treatment: an overview of applications. *Chemosphere* 173: 474–484.

Luderitz, V. and Gerlach, F. (2002). Phosphorus removal in different constructed wetlands. *Acta Biotechnology* 22 (1–2): 91–99.

Ma, J., Lenhart, J.H., and Karel, T. (2011). Orthophosphate adsorption equilibrium and breakthrough on filtration media for storm-water runoff treatment. *Journal of Irrigation and Drainage Engineering* 137 (4): 244–250.

MacLean, C. (2017). 5-year fight removes less than 1% of phosphorus from Lake Winnipeg basin. https://www.cbc.ca/news/canada/manitoba/lake-winnipeg-phosphorus-algae-blooms-1.4293366 (accessed 27 April 2019).

Mann, R. (1997). Phosphorus adsorption and desorption characteristics of constructed wetland gravels and steelworks byproducts. *Australian Journal of Soil Resources* 35: 375–384.

May, L., Place, C., O'Malley, M. et al. (2010). The impact of phosphorus inputs from small discharges on designated freshwater sites. Final report to Natural England and Broads Authority. http://publications.naturalengland.org.uk/publication/6150557569908736 (accessed 27 April 2019).

May, L., Withers, P., Jarvie, H. et al. (2014). The impact of onsite sewage treatment systems on river quality in UK river catchments. http://www.epa-pictaural.com/s/wwater12/lindaMay.php?playVideo=true (accessed 27 April 2019).

McDowell, R.W., Sharpley, A.N., and Bourke, W. (2008). Treatment of drainage water with industrial by-products to prevent phosphorus loss from tile-drained land. *Journal of Environmental Quality* 37: 1575–1582.

Mehr, J., Jedelhauser, M., and Binder, C.R. (2018). Transition of the Swiss phosphorus system towards a circular economy—part 1: current state and historical developments. *Sustainability* 10: 1479–1496. https://doi.org/10.3390/su10051479.

Merkourakis, S., Calleja, I., Delgado, L. et al. (2007). *Environmental Technologies Verification Systems. JRC European Commission Report EUR 22933 EN - 2007*. Seville: Institute for Prospective Technological Studies 112 pp.

Michaud, A.R., Ruyet, F., and Beaudin, I. (2009). Évaluation des outils de gestion agroenvironnementale à l'échelle du bassin versant dans un cadre opérationnel de serviceconseil à la ferme. Dans le cadre du projet Lisière verte. 63 pp. https://irda.blob.core.windows.net/media/2335/michaud-et-al-2009_rapport_lisiere_verte_evaluation_outils.pdf (accessed 12 May 2019).

Mihelcic, J.R., Fry, L.M., and Shaw, R. (2011). Global potential of phosphorus recovery from human urine and feces. *Chemosphere* 84 (6): 832–839.

Miller, E.C. and Wright, W.S. (2014). GeoEngineers, Inc.; April 9, 2014; Spokane County Shoreline Management Plan (SMP); Proposed Elements of a Phosphorus Management Standard. File No. 0188-163-01 Washington: City of Spokane.

Minton, G.R. and Carlson, D.A. (1972). Combined biological-chemical phosphorus removal. *Journal (Water Pollution Control Federation)* 44 (9): 1736–1755. https://www.jstor.org/stable/25037603 (accessed 27 April 2019).

Mnthambala, F., Maida, J.H.A., Lowole, M.W. et al. (2015). Phosphorus sorption and external phosphorus requirements of ultisols and oxisols in Malawi. *Journal of Soil Science and Environmental Management* 6 (3): 35–41.

Mollah, M.Y.A., Morkovsky, P., Gomes, J.A.G. et al. (2004). Fundamentals, present and future perspectives of electrocoagulation. *Journal of Hazardous Materials* 114 (1–3): 199–210.

Mollah, M.Y.A., Schennach, R., Parga, J.R. et al. (2001). Electrocoagulation (EC)—science and applications. *Journal of Hazardous Materials* B84: 29–41.

Molle, P., Lienard, A., Grasmick, A. et al. (2003). Phosphorus retention in subsurface constructed wetlands: investigations focused on calcareous materials and their chemical reactions. *Water Science and Technology* 48 (5): 75–83.

Monfreda, C., Ramankutty, N., and Foley, J.A. (2008). Farming the planet: 2. Geographic distribution of crop areas, yields, physiological types, and net primary production in the year 2000. *Global Biogeochemical Cycles* 22: GB1022. https://doi.org/10.1029/2007GB002947.

Moro, M.A., McKnight, U.S., Smets, B.F. et al. (2018). The industrial dynamics of water innovation: a comparison between China and Europe. *International Journal of Innovation Studies* 2: 14–32.

Morse, G.K., Brett, S.W., Guy, J.A. et al. (1998). Review: phosphorus removal and recovery technologies. *Science of the Total Environment* 212 (1): 69–81.

Moss, B., Kosten, S., Meerhoff, M. et al. (2011). Allied attack: climate change and eutrophication. *Inland Waters* 1: 101–105. https://doi.org/10.5268/IW-1.2.359.

Munir, M. (2013). History and evolution of the polluter pays principle: how an economic idea became a legal principle? https://papers.ssrn.com/sol3/papers.cfm?abstract_id=2322485 (accessed 27 April 2019).

Murty, M.N. and Kumar, S. (2011). *Water Pollution in India: An Economic Appraisal.* India infrastructure report 19, 285–298. New Delhi: Oxford University Press http://www.idfc.com/pdf/report/IIR-2011.pdf (accessed 27 April 2019).

Nair, P.S., Logan, T.J., Sharpley, A.N. et al. (1984). Interlaboratory comparison of a standardized phosphorus adsorption procedure. *Journal of Environmental Quality* 13 (4): 591–595.

National Centers for Coastal Ocean Science (NCCOS) (2018). Harmful algal bloom and hypoxia research and control act. https://coastalscience.noaa.gov/research/stressor-impacts-mitigation/habhrca (accessed 27 April 2019).

National Sanitation Foundation (NSF) International (2018). NSF. The Public Health and Safety Organization www.nsf.org (accessed 27 April 2019).

National Toxicology Program (2017). Microcystin TOXICITY. US Department of Health and Human Services. https://ntp.niehs.nih.gov/ntp/htdocs/chem_background/exsumpdf/microcystin_508.pdf, (accessed 13 August 2017).

Neal, C., Jarvie, H.P., Withers, P.J.A. et al. (2010). The strategic significance of waste-water sources to pollutant phosphorus levels in English rivers and to environmental management for rural, agricultural and urban catchments. *Science of the Total Environment* 408: 1485–1500.

Neall, V.E. (2013). Volcanic soils. Land use, land cover and soil sciences, vol VII. In: Encyclopedia of Life Support Systems (EOLSS), Paris: UNESCO. http://www.eolss.net/sample-chapters/c19/E1-05-07-13.pdf (accessed 27 April 2019).

Nelson, B.M. (2008). Water reform in Brazil: an analysis of its implementation in the Paraíba do Sul Basin and a consideration of social marketing as a tool for its optimal success. MSc thesis. University of Michigan, August, 2008.

Neumann, S. and Fatula, P. (2009). Principles of ion exchange in wastewater treatment. Techno Focus. https://www.scribd.com/document/327738757/090316-asian-water-principles-of-ion-exchange-neumann-03-09-pdf (accessed 27 April 2019).

New Jersey Corporation for Advanced Technology, NJCAT (2018). Technology verification. http://www.njcat.org/ (accessed 17 December 2018)

Nguyen, D.D., Ngo, H.H., Guo, W. et al. (2016). Can electrocoagulation process be an appropriate technology for phosphorus removal from municipal wastewater? *Science of the Total Environment* 563–564: 549–556.

Nilsson, C., Renman, G., Johansson Westholm, B. et al. (2013). Effect of organic load on phosphorus and bacteria removal from wastewater using alkaline filter materials. *Water Research* 47 (16): 6289–6297.

Novotny, V. and D'Arcy, B. (eds.) (1999). Diffuse pollution '98. In: *Proceedings of the IAWQ 3rd International Conference on Diffuse Pollution* Edinburgh, UK, 31 August–4 September 1998. London: IWA Publishing 367 pp.

NSW Department of Local Government (2000). *The Easy Septic Guide*. Nowra: Developed by Social Change Media for the New South Wales Department of Local Government.

NuReSys (2019). NuReSys applications. http://www.nuresys.be/technology.html (accessed 8 January 2019).

OECD (1982). *Eutrophication of Waters: Monitoring, Assessment and Control*. Paris: Organization for Economic and Cooperation and Development. 154 pp. ISBN: 9264122982.

OECD (1992). *The Polluter-Pays Principle. OECD Analyses and Recommendations*. Paris: OECD Environment Directorate. 49 pp. http://www.oecd.org/officialdocuments/publicdisplaydocumentpdf/?cote=OCDE/GD(92)81&docLanguage=En (accessed 27 April 2019).

OECD (2011). Better policies to support eco-innovation. OECD Studies on Environmental Innovation. https://www.oecd.org/env/consumption-innovation/47347714.pdf (accessed 27 April 2019).

OECD (2012). *Water quality and agriculture. Meeting the Policy Challenge*, OECD Studies on Water. Paris: OECD Publishing, 156 pp. https://www.oecd.org/publications/water-quality-and-agriculture-9789264168060-en.htm (accessed 27 April 2019).

OECD (2013). *OECD Project on Environmental Policy and Technological Innovation*. Paris: OECD Environment Directorate http://www.oecd.org/env/consumption-innovation/Brochure%206%2003%202013.pdf (accessed 27 April 2019).

OECD (2018a). Regulatory reform. Regulatory Reform and Innovation. https://www.oecd.org/sti/inno/2102514.pdf (accessed 15 April 2018).

OECD (2018b). Environmental policy and technological innovation (EPTI). http://www.oecd.org/env/consumption-innovation/innovation.htm (accessed 27 April 2019).

Oehmen, A., Lemos, P.C., Carvalho, G. et al. (2007). Advances in enhanced biological phosphorus removal: from micro to macro scale. *Water Research* 41: 2271–2300.

Ohio EPA (2013). *Cost Estimate of Phosphorus Removal at Wastewater Treatment Plants*. Columbus: Ohio Environmental Protection Agency.

Ohtake, H. and Tsuneda, S. (eds.) (2019). *Phosphorus Recovery and Recycling*. Dordrecht: Springer, 526 pp. ISBN: 978-981-10-8030-2 https://doi.org/10.1007/978-981-10-8031-9.

Olsen, S.R. and Watanabe, F.S. (1957). A method to determine a phosphorus adsorption maximum of soils as measured by the langmuir isotherm. *Soil Science Society of America Journal* 21 (2): 144–149.

Olson-Sawyer, K. (2017). Waterkeepers fighting harmful algal blooms from coast to coast. https://waterkeeper.org/waterkeepers-fighting-harmful-algal-blooms-from-coast-to-coast (accessed 27 April 2019).

OMAFRA (2015). Best Management Practices. Guelph: Ontario Ministry of Agriculture, Food and Rural Affairs. http://www.omafra.gov.on.ca/english/environment/bmp/series.htm (accessed 27 April 2019).

O'Neil, J.M., Davis, T.W., Burford, M.A. et al. (2012). The rise of harmful cyanobacteria blooms: the potential roles of eutrophication and climate change. *Harmful Algae* 14: 313–334. Special issue: Harmful Algae – The requirement for species-specific information.

Optiroc Group AB (2003). Product specification of Filtralite P, Oslo, Norway, 2003.

Ostara (2019a). Nutrient management solutions. www.ostara.com (accessed 6 January 2019).

Ostara (2019b). Metropolitan water reclamation district of Greater Chicago and Ostara open world's largest nutrient recovery facility to help recover phosphorus and protect Mississippi River Basin. http://ostara.com/project/metropolitan-water-reclamation-district-greater-chicago (accessed 6 January 2019).

Outotech (2019a). Sludge is biofuel. https://www.outotec.com/company/about-outotec/rd-and-innovation/EU-Life-Project/ (accessed 9 January 2019).

Outotech (2019b). Closing the Global Nutrient Loop (CLOOP). http://www.outotec.com/company/about-outotec/rd-and-innovation/cloop/ (accessed 10 January 2019).

Outotech News (2011). Outotec acquired phosphorus recycling technology business of ASH DEC Umwelt AG. http://www.outotec.com/company/media/news/2011/outotec-acquired-phosphorus-recycling-technology-business-of-ash-dec-umwelt-ag (accessed 27 April 2019).

Outotech News (2016). Sustainable sewage sludge incineration for Zürich canton. https://www.outotec.com/company/media/news/2016/sustainable-sewage-sludge-incineration-for-zurich-canton (accessed 27 April 2019).

Paerl, H.W. and Paul, V.J. (2012). Climate change: links to global expansion of harmful Cyanobacteria. *Water Research* 46: 1349–1363.

Panagopoulos, Y., Makropoulos, C., and Mimikou, M. (2011). Reducing surface water pollution through the assessment of the cost-effectiveness of BMPs at different spatial scales. *Journal of Environmental Management* 92 (10): 2823–2835.

Parish, L. (2008). GIS mapping of surface discharge septic systems. https://www.ilwaterconference.org/uploads/5/8/3/0/58302019/parish.pdf.

Pearce, B.J. (2015). Phosphorus Recovery Transition Tool (PRTT): a transdisciplinary framework for implementing a regenerative urban phosphorus cycle. *Journal of Cleaner Production* 109: 203–215.

Pearce, F. (2011). *Phosphate: A Critical Resource Misused and Now Running Low*. Yale Environment 360. New Haven: The Yale School of Forestry and Environmental Studies. https://e360.yale.edu/features/phosphate_a_critical_resource_misused_and_now_running_out (accessed 27 April 2019).

Pearl, H. and Huisman, J. (2008). Blooms like it hot. *Science* 320 (5872): 57–58. https://doi.org/10.1126/science.1155398.

Penn, C.J. and Bowen, J.M. (2018). *Design and Construction of Phosphorus Removal Structures for Improving Water Quality*. Cham: Springer International Publishing: 228 pp. ISBN: 978-3-319-58657-1.

Penn, C.J., Bryant, R.B., Callahan, M.P. et al. (2011). Use of industrial by-products to sorb and retain phosphorus. *Communications in Soil Science and Plant Analysis* 42 (6): 633–644.

Penn, C.J., Chagas, I., Klimeski, A. et al. (2017). A review of phosphorus removal structures: how to assess and compare their performance. *Water* 2017 (9): 583–605. https://doi.org/10.3390/w9080583.

Penn, C.J., McGrath, J.M., Rounds, E. et al. (2012). Trapping phosphorus in runoff with a phosphorus removal structure. *Journal of Environmental Quality* 41: 672–679.

Phosphorus Futures (2018). Phosphorus Sustainability. http://phosphorusfutures. net/phosphorus-sustainability (accessed 27 April 2019).

Pidou, M. (2015). Technically speaking: new approaches for phosphorus removal. *Water & Wastewater Treatment Online* (26 January): https://wwtonline.co.uk/ Error/AnoymousSubscribe?articleUrl=technically-speaking-new-approaches-for-phosphorus-approval- (accessed 3 August 2018).

Pipeline (2013). Phosphorus and onsite wastewater systems. *Pipeline* 24 (1): 1–9. http://www.nesc.wvu.edu/pdf/WW/publications/pipline/PiL_SU13.pdf (accessed 27 April 2019).

Plaxton, W.C. and Lambers, H. (eds.) (2015). *Annual Plant Reviews Volume 48: Phosphorus Metabolism in Plants*. Chichester: Willey 439 pp., doi:https://doi. org/10.1002/9781118958841.

van der Ploeg, R.R., Böhm, W., and Kirkham, M.B. (1999). On the origin of the theory of mineral nutrition of plants and the law of the minimum. *Soil Science Society of America Journal* 63: 1055–1062. https://doi.org/10.2136/sssaj1999.6351055x.

Polhamus, M. (2018). 'Lake in crisis' targeted with new legislation. https://vtdigger. org/2018/02/09/lake-crisis-targeted-new-legislation/ (accessed 27 April 2019).

Polyakov, V., Fares, A., and Ryder, M. (2005). Precision riparian buffers for the control of nonpoint source pollutant loading into surface water: a review. *Environmental Reviews* 13: 129–144. https://doi.org/10.1139/a05-010.

Prabesh, K.C. (2018). *New Opportunities of Nutrient Recycling in Water Services*. Hämeenlinna: Häme University of Applied Sciences. ISBN: 978-951-784-792-6.

Premier, R. and Ledger, S. (2006). Good agricultural practices in Australia and Southeast Asia. *Horticultural Technology* 16 (4): 552–555.

PremierTech Aqua (2018a). Phosphorus removal and disinfection systems. https:// www.premiertechaqua.com/media/67463/flyer-dpec-can-usa.pdf (accessed 27 April 2019).

PremierTech Aqua (2018b). Ecoflo biofilter – treatment and polishing unit. https:// www.premiertechaqua.com/wastewater-sewer-treatment-plants/biofilter-disinfection-peat (accessed 27 April 2019).

Pretty, J., Mason, C., Nedwell, D.B. et al. (2003). Policy analysis environmental costs of freshwater eutrophication in England and Wales. *Environmental Science & Technology* 37 (2): 201–208.

Proctor, D.M., Fehling, K.A., Shay, E.C. et al. (2000). Physical and chemical characteristics of blast furnace, basic oxygen furnace, and electric arc furnace steel industry slags. *Environmental Science & Technology* 34: 1576–1582.

Pulkka, S., Martikainen, M., Bhatnagar, A. et al. (2014). Electrochemical methods for the removal of anionic contaminants from water – a review. *Separation and Purification Technology* 132: 252–271.

Ramasahayam, S.K., Guzman, L., Gunawan, G. et al. (2014). A comprehensive review of phosphorus removal technologies and processes. *Journal of Macromolecular Science, Part A: Pure and Applied Chemistry* 51: 538–545.

Randhir, T.O., Lee, J.G., and Engel, B. (2000). Multiple criteria dynamic spatial optimization to manage water quality on a watershed scale. *Transactions of the ASAE* 43 (2): 291–299.

Rao, N.S., Easton, Z.M., Schneiderman, E.M. et al. (2009). Modeling watershed-scale effectiveness of agricultural best management practices to reduce phosphorus loading. *Journal of Environmental Management* 90 (3): 1385–1395.

Rapf, M., Raupenstrauch, H., Cimatoribus, C. et al. (2012). A new thermo-chemical approach for the recovery of phosphorus from sewage sludge. In: *Proceedings of the 11th DepoTech Conference*, Leoben, Austria, 6–9 November 2012, 691–697. http://www.vivis.de/phocadownload/Download/2012_wm/2012_WM_691_698_Rapf.pdf (accessed 27 April 2019).

Ready, M. (2008). Top reasons for septic system failure and how to prevent them. http://fremontcountywy.org/wp-content/uploads/2009/11/WhySepticSystems Fail.pdf (accessed 27 April 2019).

Reddy, K.R., Fisher, M.M., Wang, Y. et al. (2007). Potential effects of sediment dredging on internal phosphorus loading in a shallow, subtropical Lake. *Lake and Reservoir Management* 23 (1): 27–38. https://doi.org/10.1080/07438140709353907.

Renman, A. (1998). Onsite wastewater treatment – polonite and other filter materials for removal of metals, nitrogen and phosphorus. PhD thesis. KTH, Stockholm, June 2008.

Renman, A. and Renman, G. (2010). Long-term phosphate removal by the calcium-silicate material Polonite in wastewater filtration systems. *Chemosphere* 79 (6): 659–664.

Renman, G. and Thilander, O.H. (2006). Method and device for purification of wastewater. Swedish Patent PCT/EP2006/004548. http://www.google.com/patents/WO2007131522A2 (accessed 27 April 2019).

Restoration and Recovery (2014). Proprietary Systems: Illuminating the Vaults Beneath Our Feet. http://rrstormwater.com/proprietary-systems-illuminating-vaults-beneath-our-feet (accessed 27 April 2019).

Reuters Staff (2014). China to spend $330 billion to fight water pollution – paper. https://www.reuters.com/article/us-china-water-pollution/china-to-spend-330-billion-to-fight-water-pollution-paper-idUSBREA1H0H120140218 (accessed 25 May 2018).

Richardson, C.J. (1985). Mechanisms controlling phosphorus retention capacity in freshwater wetlands. *Science* 228: 1424–1427.

Richardson, C.J. and Craft, C.B. (1993). Effective phosphorus retention in wetlands: fact or fiction. In: *Constructed Wetlands for Water Quality Improvement* (ed. G.A. Moshiri), 271–282. Boca Raton: Lewis Publishers.

Richardson, J.S., Naiman, R.J., and Bisson, P.A. (2012). How did fixed-width buffers become standard practice for protecting freshwaters and their riparian areas from forest harvest practices? *Freshwater Science* 31 (1): 232–238.

Rittmann, B.E., Mayer, B., Westerhoff, P. et al. (2011). Capturing the lost phosphorus. *Chemosphere* 84: 846–853.

Roberts, W.M., Stutter, M.I., and Haygarth, P.M. (2012). Phosphorus retention and remobilization in vegetated buffer strips: a review. *Journal of Environmental Quality* 41: 389–399.

Roeleveld, P., Loeffen, P., Temmink, H. et al. (2004). Dutch analysis for P-recovery from municipal wastewater. *Water Science and Technology* 49 (10): 191–199.

Ronka, M.T.H., Lennart, C., Saari, V. et al. (2005). Environmental changes and population trends of breeding waterfowl in northern Baltic Sea. *Annales Zoologici Fennici* 42 (6): 587–602.

Rosemarin, A., de Bruijne, G., and Caldwell, I. (2009). Peak phosphorus. The next inconvenient truth. *The Broker* 4 August. http://www.thebrokeronline.eu/Articles/Peak-phosphorus (accessed 27 April 2019).

Ross, G., Haghseresht, F., and Cloete, T.E. (2008). The effect of pH and anoxia on the performance of Phoslock, a phosphorus binding clay. *Harmful Algae* 7: 545–550.

Rotz, C.A., Sharpley, A.N., Satter, L.D. et al. (2002). Production and feeding strategies for phosphorus management on dairy farms. *Journal of Dairy Science* 85 (11): 3142–3153.

Rountree, G. (2014). Thoughts on restoring Chesapeake Bay water quality and how septic systems became an issue. NOAA Seminar Series, Silver Spring, MD, 15 October 2014.

Roy, E.D. (2017). Phosphorus recovery and recycling with ecological engineering: a review. *Ecological Engineering* 98: 213–227. https://doi.org/10.1016/j.ecoleng.2016.10.076.

Roy-Poirier, A., Champagne, P., and Filion, Y. (2010). Bioretention processes for phosphorus pollution control. *Environmental Reviews - NRC Research Press* 18: 159–173.

Rozema, E.R., VanderZaag, A.C., Wood, J.D. et al. (2016). Constructed wetlands for agricultural wastewater treatment in Northeastern North America: a review. *Water* 8 (5): 173. https://doi.org/10.3390/w8050173.

Russell, J.R., Kovar, J., Morrical, D.G. et al. (2006). Effects of grazing management on pasture characteristics affecting sediment and phosphorus pollution of pasture streams (progress report). Animal Industry Report: AS 652, ASL R2122. http://lib.dr.iastate.edu/ans_air/vol652/iss1/45 (accessed 27 April 2019).

Ruzhitskaya, O. and Gogina, E. (2017). Methods for removing of phosphates from wastewater. MATEC Web of Conferences 106, 07006 (2017). https://www.matec-conferences.org/articles/matecconf/pdf/2017/20/matecconf_spbw2017_07006.pdf (accessed 27 April 2019).

Sakadevan, K. and Bavor, H.J. (1998). Phosphate adsorption characteristics of soils, slags and zeolite to be used as substrates in constructed wetland systems. *Water Research* 32 (2): 393–399.

Sarvajayakesavalu, S., Lu, Y., Withers, P.J.A. et al. (2018). Phosphorus recovery: a need for an integrated approach. *Ecosystem Health and Sustainability* 4 (2): 48–57. https://doi.org/10.1080/20964129.2018.1460122.

Schaefer, K. (2014). In the fight against green slime on Lake Erie, farmers try to clean up their act. https://www.pri.org/stories/2014-10-27/fight-against-green-slime-lake-erie-farmers-try-clean-their-act (accessed 27 April 2019).

Schaum, C. (ed.) (2019). *Phosphorus Polluter and Resource of the Future – Removal and Recovery from Wastewater*. London: IWA. 590 pp. ISBN: 9781780408354.

Schindler, D.W. (1974). Eutrophication and recovery in experimental lakes: implications for lake management. *Science* 184 (4139): 897–899.

Schindler, D.W. (1977). Evolution of phosphorus limitation in lakes. *Science* 195 (4275): 260–262. https://doi.org/10.1126/science.195.4275.260.

Schindler, D.W. (2012). The dilemma of controlling cultural eutrophication of lakes. *Proceedings of the Royal Society B: Biological Sciences* https://doi.org/10.1098/rspb.2012.1032.

Schindler, D.W., Hecky, D.L., Stainton, M.P. et al. (2008). Eutrophication of lakes cannot be controlled by reducing nitrogen input: results of a 37-year whole-eco-system experiment. *Proceedings of the National Academy of Sciences of the United States of America (PNAS)* 105 (32): 11254–11258. https://doi.org/10.1073/pnas.0805108105.

Schoenberger, H., Canfora, P., Dri, M. et al. (2014). Development of the EMAS sectoral reference documents on best environmental management practice. Learning from frontrunners promoting best practice. European Commission JRC Scientific and Policy Reports. https://ec.europa.eu/jrc/en/publication/eur-scientific-and-technical-research-reports/development-emas-sectoral-reference-documents-best-environmental-management-practice (accessed 9 September 2018).

Schoumans, O.F. (2014). BIOREFINE. Phosphorus recovery from animal manure. https://www.biorefine.eu/sites/default/files/publication-uploads/biorefine_p-recovery.pdf (accessed 27 April 2019).

Schoumans, O.F., Bouraoui, F., Kabbe, C. et al. (2015). Phosphorus management in Europe in a changing world. *Ambio* 44: 180–192. http://dx.doi.org/10.1007/s13280-014-0613-9.

Schoumans, O.F., Chardon, W.J., Bechmann, M. et al. (2011). Mitigation options for reducing nutrient emissions from agriculture. A study amongst European member states of Cost action 869. Wageningen: Alterra, Wageningen-UR. Alterra report 2141. ISSN: 1566-7197, 147 pp.

Schoumans, O.F., Ehlert, P.A.I., Nelemans, J.A. et al. (2014). Explorative study of phosphorus recovery from pig slurry: laboratory experiments. Wageningen: Alterra, Wageningen-UR. Alterra report 2514. 50 pp. http://library.wur.nl/WebQuery/wurpubs/452759 (accessed 27 April 2019).

Schoumans, O. F., Rulkens, W.H., Oenema, O. et al. (2010). Phosphorus recovery from animal manure. Wageningen: Wageningen UR. Alterra report 2158. ISSN: 1566-7197.

Schröder, J.J., Cordell, D., Smit, A.L. et al. (2010). Sustainable use of phosphorus. EU Tender ENV B.1./ETU/2009/0025. Report 357. Plant Research International Wageningen UR. http://library.wur.nl/WebQuery/wurpubs/404463 (accessed 27 April 2019).

Schwamborn, R., Bonecker, S.L.C., Galvão, I.B. et al. (2004). Mesozooplankton grazing under conditions of extreme eutrophication in Guanabara Bay, Brazil. *Journal of Plankton Research* 26 (9): 983–992. https://doi.org/10.1093/plankt/fbh090.

Schwarte, K.A., Russell, J.R., Kovar, J.L. et al. (2011). Grazing management effects on sediment, phosphorus, and pathogen loading of streams in cool-season grass pastures. *Journal of Environmental Quality* 40 (4): 1303–1313.

Scope Newsletter (2013). Thames Water launches UK's first full-scale nutrient recovery unit. Scope Newsletter 99. https://phosphorusplatform.eu/images/download/ScopeNewsletter99.pdf (accessed 27 April 2019).

Scope Newsletter (2015). Scope newsletter special issue ESPC2 – 2nd European sustainable phosphorus conference. Scope Newsletter 111. https://phosphorusplatform.eu/images/scope/ScopeNewsletter_111_special_ESPC2.pdf (accessed 27 April 2019).

Scope Newsletter (2018). SCOPE 128 newsletter science special on phosphorus. https://www.phosphorusplatform.eu/activities/r-d-and-projects/1764-scope-128-newsletter-science-special (accessed 27 April 2019).

Sellner, B.M., Hua, G., Ahiablame, L.M. et al. (2017). Evaluation of industrial by-products and natural minerals for phosphate adsorption from subsurface drainage. *Environmental Technology* 4: 1–12.

Selman, M., Sugg, Z., Greenhalgh, S. et al. (2008). Eutrophication and hypoxia in coastal areas. A global assessment of the state of knowledge. http://www.wri.org/publication/eutrophication-and-hypoxia-coastal-areas (accessed 8 June 2010).

Seo, Y.I., Hong, K.H., Kim, S.H. et al. (2013). Phosphorus removal from wastewater by ionic exchange using a surface-modified Al alloy filter. *Journal of Industrial and Engineering Chemistry* 19 (3): 744–747.

SERA 17 (2018). SERA 17 – Innovative solutions to minimize phosphorus losses from agriculture. www.sera17.org (accessed 5 September 2018).

Sharpley, A. (2016). Managing agricultural phosphorus to minimize water quality impacts. *Scientia Agricola* 73 (1): 1–8. http://dx.doi.org/10.1590/0103-9016-2015-0107.

Sharpley, A.N., Bergstrom, L., Aronsson, H. et al. (2015). Future agriculture with minimized phosphorus losses to waters: research needs and direction. *Ambio* 44 (Suppl 2): 163–179. https://doi.org/10.1007/s13280-014-0612-x.

Sharpley, A.N., Chapra, S.C., Wedepohl, R. et al. (1994). Managing agricultural phosphorus for protection of surface waters: issues and options. *Journal of Environmental Quality* 23: 437–451.

Sharpley, A.N., Daniel, T., Gibson, G. et al. (2006). Best management practices to minimize agricultural phosphorus impacts on water quality. USDA ARS-163, 52 pp. https://sera17dotorg.files.wordpress.com/2015/02/bmps-for-p-ars-163-2006.pdf (accessed 27 April 2019).

Sharpley, A.N., Kleinman, P.J.A., Flaten, D.N. et al. (2011). Critical source area management of agricultural phosphorus: experiences, challenges and opportunities. *Water Science and Technology* 64 (4): 945–952.

Sharpley, A.N., Kleinman, P.J.A., Jordan, P. et al. (2009). Evaluating the success of phosphorus management from field to watershed. *Journal of Environmental Quality* 38: 1981–1988.

Shokouhi, A. (2017). Phosphorus removal from wastewater through struvite precipitation. Master thesis 2017. Norwegian University of Life Sciences. Faculty of Environmental Sciences and Natural Resource Management. 74 pp.

Shortle, J., Kaufman, Z., Abler, D. et al. (2013). Building capacity to analyze the economic impacts of nutrient trading and other policy approaches for reducing agriculture's nutrient discharge into the Chesapeake Bay Watershed. Final Report, prepared for the USDA Office of the Chief Economist for the Cooperative Agreement No. 58-0111-11-006, August 2013. https://www.usda.gov/oce/environmental_markets/files/EconomicTradingCBay.pdf (accessed 27 April 2019).

Sims, J.T., Bergström, L., Bowman, B.T. et al. (2005). Nutrient management for intensive animal agriculture: policies and practices for sustainability. *Soil Use and Management* 21 (1): 141–151.

Singh, K.M., Singh, R.K.P., Meena, M.S. et al. (2015). Indian national water policy: a review. In: *Water Management in Agriculture* (eds. M.S. Meena, K.M. Singh and B.P. Bhatt), 47–64. Delhi: Jaya Publishing House.

Smeltzer, E. (2013). History of the Lake Champlain phosphorus TMDL. http://www.emcenter.org/wp-content/uploads/2013/02/History-of-Lake-Champlain-T.M.D.L.pdf (accessed 27 April 2019).

Smith, B. (2015). Brazil: environmental issues, policies and clean technology. *Azocleantech* https://www.azocleantech.com/article.aspx?ArticleID=547 (accessed 27 April 2019).

Smith, V.H. and Schindler, D.W. (2009). Eutrophication science: where do we go from here? *Trends in Ecology & Evolution* 24: 201–207.

Spears, B.M., Gunn, I., Andrews, C. et al. (2014). Annual review of chemical and ecological responses in Hatchmere and Mere Mere following Phoslock® application – 2013. Environment Agency Report Number SC120064/R3, 25 March.

Srivastava, P., Hamlett, J.M., Robillard, P.D. et al. (2002). Watershed optimization of best management practices using AnnAGNPS and a genetic algorithm. *Water Resources Research* 38 (3): 1–14.

State Onsite Regulators Alliance (SORA) (2012). *Onsite Wastewater Nutrient Regulation Survey Report*. Morgantown: National Environmental Services Center.

Statista (2018a). *Number of Dogs in the United States from 2000 to 2017 (in millions)*. The Statistics Portal https://www.statista.com/statistics/198100/dogs-in-the-united-states-since-2000/ (accessed 27 April 2019).

Statista (2018b). Cattle population worldwide. https://www.statista.com/statistics/263979/global-cattle-population-since-1990 (accessed 10 November 2018).

Steen, I. (1998). Phosphorus availability in 21st century: management of a non-renewable resource. *Phosphorus and Potassium* 217: 25–31.

Stensel, H.D. (1991). Principles of biological phosphorus removal. In: *Phosphorus and Nitrogen Removal from Municipal Wastewater*, 2ee (ed. R.I. Sedlak), 141–166. New York: Lewis Publishers.

Stutter, M.I., Chardon, W.J., and Kronvang, B. (2012). Riparian buffer strips as a multifunctional management tool in agricultural landscapes: introduction. *Journal of Environmental Quality* 41: 297–303. https://doi.org/10.2134/jeq2011.0439.

Subirats, J., Timoner, X., Sànchez-Melsió, A. et al. (2018). Emerging contaminants and nutrients synergistically affect the spread of class 1 integron-integrase (intI1) and sul1 genes within stable streambed bacterial communities. *Water Research* https://doi.org/10.1016/j.watres.2018.03.025.

Sun, C. and Wu, H. (2013). Assessment of pollution from livestock and poultry breeding in China. *International Journal of Environmental Studies* 70 (2): 232–240.

SWAT (2018). SWAT literature data base for peer reviewed journal articles https://www.card.iastate.edu/swat_articles (accessed 27 April 2019).

Swedish EPA 2006. Naturvårdsverkets allmänna råd [till 2 och 26 kap. miljöbalken och 12-14 och 19 §§ förordningen (1998:899) om miljöfarlig verksamhet och hälsoskydd] om små avloppsanordningar för hushållsspillvatten. Naturvärdsverket.

Szögi, A.A., Vanotti, M.B., and Hunt, P.G. (2015). Phosphorus recovery from pig manure solids prior to land application. *Journal of Environmental Management* 157: 1–7. https://doi.org/10.1016/j.jenvman.2015.04.010.

Tait, J. and Banda, G. (2016). Proportionate and adaptive governance of innovative technologies. The role of regulations, guidelines and standards. Edinburgh: British Standards Institution.

Talberth, J., Selman, M., Walker, S. et al. (2015). Pay for performance: optimizing public investments in agricultural best management practices in the Chesapeake Bay Watershed. *Ecological Economics* 118 (2015): 252–261.

Tamini, L.D. (2009): Agri-environment advisory activities effects on best management practices adoption. https://mpra.ub.uni-muenchen.de/21955/1/MPRA_paper_21955.pdf (accessed 27 April 2019).

Tanner, C.C., Sukias, J.P.S., and Yates, C.R. (2010). *New Zealand Guidelines: Constructed Wetland Treatment of Tile Drainage*, NIWA Information Series No. 75. Auckland: National Institute of Water & Atmospheric Research Ltd.

Tarayre, C., De Clercq, L., Charlier, R. et al. (2016). New perspectives for the design of sustainable bioprocesses for phosphorus recovery from waste. *Bioresource Technology* 206: 264–274. ISSN: 1873-2976. https://doi.org/10.1016/j.biortech.2016.01.091.

Tchobanoglous, G., Burton, F.L., and Stensel, H.D. (2003). *Wastewater Engineering: Treatment and Reuse* (International Edition). New Delhi: Tata McGraw-Hill Publishing Company Ltd.

Teenstra, E., Vellinga, T., Aektasaeng, N. et al. (2014). *Global Assessment of Manure Management Policies and Practices*. Wageningen: Wageningen UR Livestock Research. 35 pp. http://www.wageningenur.nl/upload_mm/a/2/f/8a7d1a1e-2535-432b-bab5-fd10ff49a2b1_Global-Assessment-Manure-Management.pdf (accessed 27 April 2019).

The Economist (2011). Global livestock counts. Counting Chickens. The Economist online, 27 July 2011. http://www.economist.com/blogs/dailychart/2011/07/global-livestock-counts (accessed 1 August 2015).

*The Guardian* (2014). Value from sewage? A new technology cleans up waste water. *The Guardian*: 7 March https://www.theguardian.com/sustainable-business/value-sewage-technology-clean-waste-water-ostara (accessed 27 April 2019).

The White House (2009). Executive Order 13508-- Chesapeake Bay Protection and Restoration. Washington: Office of the Press Secretary. https://obamawhitehouse.archives.gov/realitycheck/the_press_office/Executive-Order-Chesapeake-Bay-Protection-and-Restoration (accessed 27 April 2019).

Tian, Y., He, W., Zhu, X. et al. (2017). Improved electrocoagulation reactor for rapid removal of phosphate from wastewater. *ACS Sustainable Chemistry Engineering* 5: 67–71. https://doi.org/10.1021/acssuschemeng.6b01613.

Tibbetts, J. (2005). Combined sewer systems: down, dirty, and out of date. *Environmental Health Perspectives* 113 (7): A464–A467. https://www.ncbi.nlm.nih.gov/pmc/articles/PMC1257666/pdf/ehp0113-a00464.pdf (accessed 27 April 2019).

Tomer, M.D., James, D.E., and Isenhart, T.M. (2003). Optimizing the placement of riparian practices in a watershed using terrain analysis. *Journal of Soil Water Conservation* 58 (4): 198–206.

Toor, G.S., Lusk, M., and Obreza, T. (2011). *Onsite Sewage Treatment and Disposal Systems: An Overview*. Gainesville: University of Florida – IFAS http://edis.ifas.ufl.edu/ss549 (accessed 29 June 2015).

Topare, N.S., Attar, S.J., and Manfe, M.M. (2011). Sewage/wastewater treatment technologies: a review. *Scientific Reviews & Chemical Communications* 1 (1): 18–24. http://www.tsijournals.com/articles/sewagewastewater-treatment-technologies-a-review.pdf (accessed 27 April 2019).

Torbert, H.A., King, K.W., and Harmel, R.D. (2005). Impact of soil amendments on reducing phosphorus losses from runoff in sod. *Journal of Environmental Quality* 34: 1415–1421.

Tunçsiper, B., Drizo, A., and Twohig, E. (2015). Constructed wetlands as a potential management practice for cold climate dairy effluent treatment – VT, USA. *CATENA* 135: 184–192.

Ueno, Y. and Fujii, M. (2001). Three year experience in operating and selling recovered struvite from full scale plant. *Environmental Technology* 22: 1373–1381.

UKWIR (2017). CIP update and the phosphorus element. UKWIR News 82 (March) 3.

UN Water (2017). 2017 UN world water development report, wastewater: the untapped resource. http://www.unwater.org/publications/world-water-development-report-2017/ (accessed 27 April 2019).

UNEP (2010). Building the foundations for sustainable nutrient management. A publication of the Global Partnership on Nutrient Management (GPNM). http://www.ais.unwater.org/ais/pluginfile.php/225/mod_label/intro/Building_the_foundations-2.pdf (accessed 27 April 2019).

UNEP (2012). Nutrient Management BMP Summary 2012. Linking Food Security and Agriculture Production to Conservation Practices. United Nations environmental programme, global programme of action for the protection of the marine environment from land based activities. http://wedocs.unep.org/bitstream/handle/20.500.11822/10710/NutrientManagementBMPSummary.pdf?sequence=1&isAllowed=y (accessed 27 April 2019).

UNEP (2016). A snapshot of the world's water quality: towards a global assessment. https://uneplive.unep.org/media/docs/assessments/unep_wwqa_report_web.pdf (accessed 27 April 2019).

UNESCO (2017). Brazil's sectorial funds on a mission to boost innovation. Science, Technology and Innovation Policy. Natural Sciences Sector May 23. http://www.unesco.org/new/en/natural-sciences/science-technology/single-view-sc-policy/news/brazils_sectorial_funds_face_challenges_in_their_mission/ (accessed 27 April 2019).

United Nations Environmental Programme (2017). Water quality: the impact of eutrophication. Newsletter and Technical Publications. Lakes and Reservoirs vol. 3. http://www.unep.or.jp/ietc/publications/short_series/lakereservoirs-3/1.asp (accessed 6 August 2017).

United States Government (2014). Harmful Algal Bloom and Hypoxia Research and Control Amendments Act of 2014. https://www.congress.gov/113/plaws/publ124/PLAW-113publ124.pdf (accessed 27 April 2019).

United States Government (2015). Safe Drinking Water Act. https://www.gpo.gov/fdsys/pkg/PLAW-114publ45/html/PLAW-114publ45.htm (accessed 27 April 2019).

United States Government (2017). Harmful Algal Bloom and Hypoxia Research and Control Amendments Act of 2017. https://www.congress.gov/bill/115th-congress/house-bill/4417 (accessed 27 April 2019).

United States Government Printing Office (1987). PUBLIC LAW 100-4—FEB. 4, 1987. https://www.gpo.gov/fdsys/pkg/STATUTE-101/pdf/STATUTE-101-Pg7.pdf (accessed 27 April 2019).

United States Goverrnment (1998). Harmful Algal Bloom and Hypoxia Research and Control Act of 1998. Public Law 105-383. November 13, 1998 112 Stat 3447. https://cdn.coastalscience.noaa.gov/page-attachments/research/habhrca.pdf (accessed 27 April 2019).

University of British Columbia (2012). Top Canadian, Indian institutions form $30M partnership to improve water and infrastructure safety, eradicate diseases. Press release. https://news.ubc.ca/2012/11/06/top-canadian-indian-institutions-form-30m-partnership-to-improve-water-and-infrastructure-safety-eradicate-diseases/ (accessed 27 April 2019).

University of Waterloo (2018). PHOSPHEX™ – treatment and removal of phosphorous from water. https://uwaterloo.ca/research/waterloo-commercialization-office-watco/business-opportunities-industry/phosphextm-treatment-and-removal-phosphorous-water (accessed 27 April 2019).

US EPA (2000a). Nutrient criteria technical guidance manual rivers and streams. EPA-822-B-00-002, July 2000. http://www.mostreamteam.org/Documents/Research/WaterQuality/EPA%20Nutrient%20Criteria%20Streams.pdf (accessed 27 April 2019).

US EPA (2000b). Wastewater technology fact sheet chemical precipitation. EPA 832-F-00-018. September 2000. https://nepis.epa.gov/Exe/ZyNET.EXE?ZyActionL=Register&User=anonymous&Password=anonymous&Client=EPA&Init=1 (accessed 27 April 2019).

US EPA (2001). Development and adoption of nutrient criteria into water quality standards. Memorandum WQSP-01-01. https://www.freshlawblog.com/files/2019/02/2001-EPA-Memo-re-Development-and-Adoption-of-Nutrient-Criteria.pdf (accessed 27 April 2019).

US EPA (2002). Onsite wastewater treatment systems manual. Office of Water, Office of Research and Development, Environmental Protection Agency EPA/625/R-00/008, February 2002. Washington: EPA.

US EPA (2003a). National pollutant discharge elimination system permit regulations and effluent limitation guidelines and standards for Concentrated Animal Feeding Operations (CAFOs); Final Rule. US Environmental Protection Agency (EPA) FRL-7424-7. Washington: EPA.

US EPA (2003b). Producers' compliance guide for CAFO's – Revised CWA Regulations for CAFO. US Environmental Protection Agency (EPA) 821-R-03-010, N. Washington: EPA.

US EPA (2004). Report to congress: impacts and control of CSOs and SSOs. August 2004. Document No. EPA-833-R-04-001. Washington: EPA.

US EPA. (2006). Revised national pollutant discharge and effluent limitation guidelines for concentrated animal feeding operations in response to waterkeeper decision; Proposed Rule. U.S. Environmental Protection Agency (EPA) – HQ-OW-2005-0037; FRL-8189-7. Washington: EPA.

US EPA (2009). Nutrient control design manual. State of Technology Review Report. EPA/600/R-09/012 (January 2009). https://nepis.epa.gov/Exe/ZyPDF.cgi/P1002X49.PDF?Dockey=P1002X49.PDF (accessed 27 April 2019).

US EPA (2011). Memorandum: working in partnership with states to address phosphorus and nitrogen pollution through use of a framework for state nutrient reductions. Published March 16th, 2011. https://www.epa.gov/sites/production/files/documents/memo_nitrogen_framework.pdf (accessed 27 April 2019).

US EPA (2013). The national rivers and streams assessment 2008-2009: a collaborative survey. http://water.epa.gov/type/rsl/monitoring/riverssurvey/upload/NRSA 200809_FactSheet_Report_508Compliant_130314.pdf (accessed 27 April 2019).

US EPA (2015). Estimated animal agriculture nitrogen and phosphorus from manure. http://www2.epa.gov/nutrient-policy-data/estimated-animal-agriculture-nitrogen-and-phosphorus-manure (accessed 27 April 2019).

US EPA (2016a). Phosphorus TMDLs for Vermont segments of Lake Champlain. 17 June 2016. https://www.epa.gov/sites/production/files/2016-06/documents/phosphorus-tmdls-vermont-segments-lake-champlain-jun-17-2016.pdf (accessed 27 April 2019).

US EPA (2016b). Memorandum: renewed call to action to reduce nutrient pollution and support for incremental actions to protect water quality and public health. Published 22 September 2016. https://www.epa.gov/sites/production/files/2016-09/documents/renewed-call-nutrient-memo-2016.pdf (accessed 27 April 2019).

US EPA (2016c). Environmental technology verification program. Archived. https://archive.epa.gov/nrmrl/archive-etv/web/html/ (accessed 27 April 2019).

US EPA (2018a). National study of nutrient removal and secondary technologies. https://www.epa.gov/eg/national-study-nutrient-removal-and-secondary-technologies#background (accessed 27 April 2019).

US EPA (2018b). Nutrient pollution policy and data. https://www.epa.gov/nutrient-policy-data (accessed 27 April 2019).

US EPA (2018c). Stormwater discharges from municipal sources. https://www.epa.gov/npdes/stormwater-discharges-municipal-sources (accessed 27 April 2019).

US EPA (2018d). Summary of the Clean Water Act 33 U.S.C. §1251 et seq (1972). https://www.epa.gov/laws-regulations/summary-clean-water-act (accessed 30 April 2018).

US EPA (2018e). Technology innovation. https://www.epa.gov/innovation/technology-innovation (accessed 27 April 2019).

US EPA (2018f). Technology innovation for environmental and economic progress. https://19january2017snapshot.epa.gov/envirofinance/innovation_.html (accessed 27 April 2019).

US Government Accountability Office (US GAO) (2016). Environmental protection Information on Federal Agencies' Expenditures and Coordination Related to Harmful Algae. Report to Congressional Committees. October 2016. https://www.gao.gov/assets/690/680457.pdf (accessed 27 April 2019).

US Office of Water (2011). Summary of state stormwater standards. https://www3.epa.gov/npdes/pubs/sw_state_summary_standards.pdf (accessed 27 April 2019).

USDA (1999). US Department of Agriculture and US Environmental Protection Agency Unified National Strategy for Animal Feeding Operations, March 9, 1999. https://www3.epa.gov/npdes/pubs/finafost.pdf (accessed 27 April 2019).

USDA (2014). USDA to invest $45 million to improve water quality in Lake Champlain. https://www.usda.gov/media/press-releases/2014/08/28/usda-invest-45-million-improve-water-quality-lake-champlain (accessed 27 April 2019).

USDA NRCS (2018). Conservation practices. http://www.nrcs.usda.gov/wps/portal/nrcs/detailfull/national/technical/cp/ncps/?cid=nrcs143_026849 (accessed 9 September 2018).

USDA NRCS Maryland (2016). Natural resources conservation service interim conservation practice standard. Phosphorus Removal System (Number) Code 782. November 2016. https://efotg.sc.egov.usda.gov/references/public/MW/MD_782_Phosphorus_Treatment_11_9_16_Draft.pdf (accessed 27 April 2019).

USDA NRCS New York (2016). Natural Resources Conservation Service interim conservation practice standard. Phosphorus Removal System (Number) Code 782. October 2016. https://efotg.sc.egov.usda.gov/references/public/NY/nyps782.pdf (accessed 27 April 2019).

USDA NRCS Pennsylvania (1995). Silage leachate and water quality. Environmental Quality Technical Note N5, 17 pp.

USDA NRCS Pennsylvania (2017). Natural resources conservation service interim conservation practice standard. Phosphorus Removal System (Number) Code 782. April, 2017. https://efotg.sc.egov.usda.gov/references/public/PA/782PhosphorusRemovalSystemPAFinalInterimStandardApril2017.pdf (accessed 27 April 2019).

USDA NRCS Vermont (2013). Natural resources conservation service interim conservation practice standard. Phosphorus Removal System (Number) Code 782. August 2013. https://efotg.sc.egov.usda.gov/references/public/VT/VT782_(1Col).pdf (accessed 27 April 2019).

USDA NRCS Wisconsin (2015). Natural resources conservation service interim conservation practice standard. Phosphorus Removal System (Number) Code 782. September 2015. https://efotg.sc.egov.usda.gov/references/public/WI/782_Standard.pdf (accessed 27 April 2019).

USGS (US Geological Survey) (1999). The Quality of our Nations Waters-Nutrients and Pesticides, US Geological Survey (USGS) Circular 1225.

USGS (US Geological Survey) (2012). Mineral commodity summaries January 2012, Phosphorus. http://minerals.usgs.gov/minerals/pubs/commodity/phosphate rock/ (accessed 10 May 2015)

Valley, L. (2016). Putting the focus on phosphorus. Utilityweek 29 February 2016. https://utilityweek.co.uk/putting-the-focus-on-phosphorus (accessed 27 April 2019).

Vanotti, M.B., Szogi, A.A., and Hunt, P.G. (2005). Wastewater treatment system. US Patent No. 6,893,567, Issued May 17, 2005. U.S. Patent & Trademark Office, Washington, DC.

Veiga, L.B.E. and Magrini, A. (2013). The Brazilian water resources management policy: fifteen years of success and challenges. *Water Resources Management* 27: 2287–2302. https://doi.org/10.1007/s11269-013-0288-1.

Veith, T.L., Wolfe, M.L., and Heatwole, C.D. (2003). Optimization procedure for cost effective BMP placement at a watershed scale. *Journal of Amercian Water Resources Association* 39 (6): 1331–1343.

Veolia (2018). ACTIFLO®. The ultimate clarification process. http://www.veoliawatertech.com/vwst-northamerica/ressources/files/1/28586,Actiflo_Industrial2014_LR-1.pdf (accessed 15 November 2018).

Vermont Agency of Natural Resources (2012). Letter issued to dr Drizo by Dr Ernest Christianson, Program Manager of Drinking Water and Ground Water Protection Division and Wastewater Disposal, regarding the State of Vermont Approval of PhosphoReduc system use for Phosphorus Removal from onsite septic systems effluents. 5 June 2012.

Vermont Agency of Natural Resources, VT ANR (2018). Department of Environmental Conservation. Laws, regulations and rules pertaining to water quality. http://dec.vermont.gov/watershed/laws (accessed 9 May 2018).

Vermont Digger (2018). Vermont Agency of Agriculture, Food & Markets Press Release 28 February 2018. Agency of Agriculture files proposed subsurface tile drainage amendment to required agricultural practices rule. https://vtdigger.org/2018/03/06/agency-agriculture-files-proposed-subsurface-tile-drainage-amendment-required-agricultural-practices-rule/ (accessed 27 April 2019).

Vermont EPSCoR (2018). Vermont EPSCoR Homepage. https://epscor.w3.uvm.edu/2 (accessed 15 November 2018).

Vermont Government (2018). Bill H.730: Conservation and development; water resources; water quality; lake in crisis. https://legislature.vermont.gov/Documents/2018/Docs/BILLS/H-0730/H-0730%20As%20passed%20by%20the%20House%20Official.pdf (accessed 27 April 2019).

Vidal, B., Hedström, A., and Herrmann, I. (2018). Phosphorus reduction in filters for on-site wastewater treatment. *Journal of Water Process Engineering* 22: 210–217.

Visvanathan, C. (2015). Status of decentralized wastewater treatment in Europe: need, drivers, trends and technologies. http://docplayer.net/15035835-Status-of-decentralized-wastewater-treatment-in-europe.html (accessed 27 April 2019).

Vohla, C., Kõiv, M., Bavor, J.H. et al. (2011). Filter materials for phosphorus removal from wastewater in treatment wetlands—a review. *Ecological Engineering* 37 (1): 70–89.

Vollenweider, R.A. (1970). Scientific fundamentals of the eutrophication of lakes and flowing waters, with particular reference to nitrogen and phosphorus as factors in eutrophication. Paris: OECD. 30th September 1970, 159 pp.

Von Sperling, M. (2016). Urban wastewater treatment in Brazil. Department of Sanitary and Environmental Engineering Federal University of Minas Gerais Brazil. Technical Note N°IDB-TN-970. https://publications.iadb.org/bitstream/handle/11319/7783/Urban-wastewater-treatment-in-Brazil.pdf (accessed 27 April 2019).

Voulvoulis, N., Arpon, K.D., and Giakoumis, T. (2017). The EU water framework directive: from great expectations to problems with implementation. *Science of the Total Environment* 575 (1): 358–366.

Vymazal, J. (2010). Constructed wetlands for wastewater treatment. *Water* 2010 (2): 530–549.

Wagner, K.J., Meringolo, D., Mitchell, D.F. et al. (2017). Aluminum treatments to control internal phosphorus loading in lakes on Cape Cod, Massachusetts. *Lake Reservoir Management* 33: 171–186.

Wakeford, J.P. (1913). Chemical precipitation at the sewage disposal works, Wakefield. *Journal of the Royal Sanitary Institute* 34 (10): 475–478. https://doi.org/10.1177/146642401303401004.

Walan, P. (2013). Modeling of peak phosphorus. a study of bottlenecks and implications for future production. Master Thesis, Uppsala University, May 2013. http://www.w-program.nu/filer/exjobb/Petter_Walan.pdf (accessed 10 May 2015).

Walker, R.L., Cotton, C.F., and Payne, K. (2003). A GIS inventory of on-site septic systems adjacent to the coastal waters of McIntosh County, Georgia. Issued by the Georgia Sea Grant College Program and the University of Georgia Marine Extension Service. Number 27, April 2003.

Walton, B. (2010). Blooming controversy: what is killing the wildlife in Kenya's Lake Naivasha? Circle of Blue Water News 16 June. http://www.circleofblue.org/2010/world/blooming-controversy-what-is-killing-the-wildlife-in-kenya%E2%80%99s-lake-naivasha/ (accessed 27 April 2019).

Wang, C. and Jiang, H. (2016). Chemicals used for in situ immobilization to reduce the internal phosphorus loading from lake sediments for eutrophication control. *Critical Reviews in Environmental Science and Technology* 46 (10): 947–997. http://dx.doi.org/10.1080/10643389.2016.1200330.

Wang, C., Liu, Z., Wang, D. et al. (2014). Effect of dietary phosphorus content on milk production and phosphorus excretion in dairy cows. *Journal of Animal Science and Biotechnology* 5: 23–29.

Wardle, T. (1893). *On Sewage Treatment and Disposal.* London: John Heywood 43pp.

Washington Stormwater Center (2018). TAPE Program. http://www.wastormwatercenter.org/tape-program (accessed 27 April 2019).

Water (2018). Special Issue 'Wetlands for the Treatment of Agricultural Drainage Water'. (ed. G. Sun). ISSN: 2073-4441. https://www.mdpi.com/journal/water/special_issues/Wetlands_Agricultural_Drainage_Water (accessed 27 April 2019).

Water and Soil Solutions International (2018). Water and Soil Solutions International website. www.phosphoreduc.com (accessed 27 April 2019).

Water Environment Federation (2014). Investigation into the feasibility of a national testing and evaluation program for stormwater products and practices. February 6, 2014 STEPP Workgroup – Steering Committee. http://wefstormwaterinstitute.org/wp-content/uploads/2016/08/WEF-STEPP-White-Paper_Final_02-06-142.pdf (accessed 27 April 2019).

Water Environment Federation (2017). The National Storm water Testing and Evaluation for Products and Practices (STEPP) Initiative. SWI-2017-FS-001, STEPP. https://wef.org/globalassets/assets-wef/direct-download-library/public/03---resources/swi-2017-fs-001-national-stepp-initiative.pdf (accessed 17 December 2018).

Water Environment Federation (WEF) (2015a). WEF releases nutrient roadmap primer. https://news.wef.org/wef-releases-nutrient-roadmap-primer (accessed 27 April 2019).

Water Environment Federation (WEF) (2015b). The nutrient roadmap. Alexandria: WEF. 184 pp. https://www.e-wef.org/Default.aspx?TabID=251&productId=45090372 (accessed 27 April 2019).

Weber, D., Drizo, A., Twohig, E. et al. (2007). Upgrading constructed wetlands phosphorus reduction from a dairy effluent using EAF steel slag filters. *Water Science and Technology* 56 (3): 135–143.

Weber, R. (2018). Lake Champlain report card: state gets a D+ for its clean-up efforts. Conservation Law Foundation, 24 April 2018. https://www.clf.org/blog/lake-champlain-report-card/ (accessed 27 April 2019).

Weber St Gobain (2018). Weber St Gobain website. https://wecare.weber/ (accessed 27 April 2019).

Wentworth, J. (2014). Phosphate resources - POST note 477, August 2014. http://www.parliament.uk/business/publications/research/briefing-papers/POST-PN-477/phosphate-resources (accessed 27 April 2019).

WERF (Water Research Foundation) (2010). Technology roadmap for sustainable wastewater treatment plants in a carbon-constrained world. https://www.werf.org/a/ka/Search/ResearchProfile.aspx?ReportId=OWSO4R07d (accessed 10 May 2015).

Wetsus (2019). Phosphate recovery. https://www.wetsus.nl/phosphate-recovery (accessed 27 April 2019).

White, J.W. (1928). *Blast Furnace Slag as a Fertilizer*, 1–19. Pennsylvania State College, School of Agriculture and Experimental Station. Bulletin 220.

Whitton, B.A. and Potts, M. (2000). *The Ecology of Cyanobacteria*. Dordrecht: Kluwer Academic Publishers.

Wik, M., Pingali, P., and Broca, S. (2008). Global agricultural performance: past trends and future prospects. Background paper for the WorldBank Development Report 2008. http://siteresources.worldbank.org/INTWDRS/Resources/477365-1327599046334/8394679-1327599874257/Pingali-Global_Agricultural_Performance.pdf (accessed 27 April 2019).

Winblad, U. and Simpson-Hebert, M. (2004). *Ecological Sanitation – Revised and Enlarged Edition*. Stockholm: Stockholm Environment Institute http://www.susana.org/en/resources/library/details/487 (accessed 27 April 2019.

Winsten, J. (2004). Developing and assessing policy options for reducing phosphorus loading in Lake Champlain. Lake Champlain Basin Program. Technical Report No. 47. http://www.lcbp.org/techreportPDF/47_P_policy_options.pdf (accessed 27 April 2019).

Withers, P.J.A. and Jarvie, H.P. (2008). Delivery and cycling of phosphorus in rivers: a review. *Science of the Total Environment* 400: 379–395.

Withers, P.J.A., Jarvie, H.P., Hodgkinson, R.A. et al. (2009). Characterization of phosphorus sources in rural watersheds. *Journal of Environmental Quality* 38: 1998–2011.

Withers, P.J.A., Jordan, P., May, L. et al. (2014a). Do septic tank systems pose a hidden threat to water quality? *Frontiers in Ecology and Environment* 12 (2): 123–130. https://doi.org/10.1890/130131.

Withers, P.J.A., May, L., Jarvie, H.P. et al. (2012). Nutrient emissions to water from septic tank systems in rural catchments: uncertainties and implications for policy. *Environmental Science and Policy* 24: 71–82.

Withers, P.J.A., Neal, C., Jarvie, H. et al. (2014b). Agriculture and Eutrophication: where do we go from here? *Sustainability* 6: 5853–5875. https://doi.org/10.3390/su6095853.

Withers, P.J.A., van Dijk, K.C., Neset, T.S. et al. (2015). Stewardship to tackle global phosphorus inefficiency: the case of Europe. *Ambio* 44 (Suppl. 2): S193–S206. https://doi.org/10.1007/s13280-014-0614-8.

World Bank (2015). Fertilizer consumption (kilograms per hectare of arable land). https://data.worldbank.org/indicator/AG.CON.FERT.ZS (accessed 27 April 2019).

World Bank (2018a). Agricultural land (sq. km). https://data.worldbank.org/indicator/AG.LND.AGRI.K2 (accessed 21 November 2018).

World Bank (2018b). Surface area (sq. km). https://data.worldbank.org/indicator/AG.SRF.TOTL.K2 (accessed 21 November 2018).

World Bank (2018c). Arable land (% of land area). https://data.worldbank.org/indicator/AG.LND.ARBL.ZS (accessed 21 November 2018).

World Bank (2018d). Fertilizer consumption (kilograms per hectare of arable land). https://data.worldbank.org/indicator/AG.CON.FERT.ZS, accessed 21 November, 2018.

World Resources Institute (WRI) (2018). Mapping, sharing data, and growing awareness on eutrophication and hypoxia around the globe. Interactive map of eutrophication and hypoxia. http://www.wri.org/our-work/project/eutrophication-and-hypoxia/interactive-map-eutrophication-hypoxia (accessed 9 January 2018).

WorldSteel Association (2018). World steel in figures 2018. https://www.worldsteel.org/en/dam/jcr:f9359dff-9546-4d6b-bed0-996201185b12/World+Steel+in+Figures+2018.pdf (accessed 27 April 2019).

Wrigley, T.J., Webb, K.M., and Venkitachalm, H. (1992). A laboratory study of struvite precipitation after digestion of piggery wastes. *Bioresource Technology* 41: 117–121.

WSDOE (2011). *Technical and Economic Evaluation of Nitrogen and Phosphorus Removal at Municipal Wastewater Treatment Facilities*. Olympia: Washington State Department of Ecology Publication No. – 11-10-060.

Wu, Y., Wen, Y., Zhou, J. et al. (2014). Phosphorus release from lake sediments: effects of pH, temperature and dissolved oxygen. *KSCE Journal of Civil Engineering* 18 (1): 323–329.

WUR (2018). Circular agriculture: a new perspective for Dutch agriculture. WUR news, published 13 September 2018. https://www.wur.nl/en/newsarticle/Circular-agriculture-a-new-perspective-for-Dutch-agriculture-1.htm (accessed 27 April 2019).

WWTonline (2016a). Phosphorus removal: how low can you go? https://wwtonline.co.uk/features/phosphorus-removal-how-low-can-you-go- (accessed 27 April 2019).

WWTonline (2016b). Severn Trent trials six technologies to reduce phosphorus. https://wwtonline.co.uk/news/severn-trent-trials-six-technologies-to-reduce-phosphorus (accessed 27 April 2019).

WWTonline (2018). Eco-friendly phosphorus removal: soneco boom. https://wwtonline.co.uk/features/eco-friendly-phosphorus-removal-soneco-boom (accessed 27 April 2019).

Wyant, K.A., Corman, J.R., and Elser, J.J. (eds.) (2013). *Phosphorus, Food and Our Future*. Oxford: Oxford University Press, 256 pp. ISBN: 9780199916832.

Xie, H., Chen, L., and Shen, Z. (2015). Assessment of agricultural best management practices using models: current issues and future perspectives. *Water* 7: 1088–1108. https://doi.org/10.3390/w7031088.

Xu, J., Luu, L., and Tang, Y. (2016). Phosphate removal using aluminum-doped magnetic nanoparticles. *Desalination and water treatment* 58: 239–248.

Xu, Z. (2016). Water policy and regulation in China. Assignment submitted for Innovative Technologies and Global Water Challenges. MSc in Water Environment Management, Heriot Watt University, November 2016.

Yamada, H., Kayama, M., Saito, K. et al. (1986). A fundamental research on phosphate removal by using slag. *Water Research* 20: 547–557.

Yao-Jen, T., Chen-Feng, Y., Chien-Kuei, C. et al. (2015). Application of magnetic nano-particles for phosphorus removal/recovery in aqueous solution. *Journal of the Taiwan Institute of Chemical Engineers* 46: 148–154.

Yoshida, H., van Dijk, K., Drizo, A. et al. (2013). Chapter 6: phosphorus recovery and reuse. In: *Phosphorus, Food and Our Future* (eds. K.A. Wyant, J.R. Corman and J.J. Elser), 112–141. Oxford: Oxford University Press.

Zarrabi, M., Soori, M.M., Sepehr, M.N. et al. (2014). Removal of phosphorus by ion exchange resins: equilibrium, kinetic and thermodynamic studies. *Environmental Engineering and Management Journal* 13 (4): 891–903.

Zhang, Y.Y., Wang, C., He, F. et al. (2016). In-situ adsorption-biological combined technology treating sediment phosphorus in all fractions. *Scientific Reports* 6: 29725. https://doi.org/10.1038/srep29725.

Zheng, X., Chen, Y., and Wu, R. (2011). Long-term effects of titanium dioxide nanoparticles on nitrogen and phosphorus removal from wastewater and bacterial community shift in activated sludge. *Environmental Science Technology* 45 (17): 7284–7290.

Zhou, T. (2017). Phosphorus recovery from wastewater and sludge: Concept for different regional conditions. Doctoral thesis. Technische Universität Berlin, Germany. https://depositonce.tu-berlin.de//handle/11303/7226 (accessed 27 April 2019).

Zhu, T., Jenssen, P.D., Mæhlum, T. et al. (1997). Phosphorus sorption and chemical characteristics of light-weight aggregates (LWA)—potential filter media in treatment wetlands. *Water Science and Technology* 35 (5): 103–108.

# Index

*Phosphorus Pollution Control - Policies and Strategies*, First Edition. Aleksandra Drizo.
© 2020 John Wiley & Sons Ltd. Published 2020 by John Wiley & Sons Ltd.